Engineering Materials

This series provides topical information on innovative, structural and functional materials and composites with applications in optical, electrical, mechanical, civil, aeronautical, medical, bio- and nano-engineering. The individual volumes are complete, comprehensive monographs covering the structure, properties, manufacturing process and applications of these materials. This multidisciplinary series is devoted to professionals, students and all those interested in the latest developments in the Materials Science field, that look for a carefully selected collection of high quality review articles on their respective field of expertise.

Indexed at Compendex (2021) and Scopus (2022)

Liwei Zhang
Editor

Corrosion in CO_2 Capture, Transportation, Geological Utilization and Storage

Causes and Mitigation Strategies

Springer

Editor
Liwei Zhang
State Key Laboratory of Geomechanics
and Geotechnical Engineering, Institute
of Rock and Soil Mechanics
Chinese Academy of Sciences
Wuhan, Hubei, China

ISSN 1612-1317 ISSN 1868-1212 (electronic)
Engineering Materials
ISBN 978-981-99-2391-5 ISBN 978-981-99-2392-2 (eBook)
https://doi.org/10.1007/978-981-99-2392-2

This Springer imprint is published by the registered company Springer Nature Singapore Pte Ltd.
The registered company address is: 152 Beach Road, #21-01/04 Gateway East, Singapore 189721, Singapore

Preface

Multiple measurements and data sets demonstrate that levels of carbon dioxide (CO_2) have been increasing in the Earth's atmosphere for the past several centuries, with the rate becoming steeper in recent decades. CO_2 is known as a "greenhouse gas" (GHG) because of its ability to trap heat in the atmosphere and raise the surface temperature of the Earth. CO_2 accounts for about 76% of the total greenhouse gas emissions, due to very significant global CO_2 emissions from various sectors. The trends toward greater concentrations of CO_2 in the atmosphere and its potential effects on climate have been a very hot research topic, and most research outcomes attribute global warming to anthropogenic CO_2 emissions.

CO_2 capture, utilization, and storage (CCUS) is a process that captures CO_2 emissions from sources like coal-fired power plants and either reuses or stores it so it will not enter the atmosphere. CCUS is a key technology to mitigate climate change and substantially reduce greenhouse gas emissions from fossil fuels. CCUS deals with high concentration CO_2, which is very corrosive in a humid environment. Corrosion can occur at all stages of CCUS, such as absorption tower corrosion induced by amine degradation products at the stage of CO_2 capture, pipeline corrosion at the stage of CO_2 transportation, wellbore casing corrosion, and wellbore cement corrosion at the stage of subsurface CO_2 utilization and storage. Therefore, it is very important to characterize, monitor, and mitigate CO_2-induced corrosion at all stages of the CCUS operation chain.

The Institute of Rock and Soil Mechanics (IRSM), Chinese Academy of Sciences has started experimental and numerical simulation works on CCUS since 2004, and our CCUS laboratory is one of the earliest CCUS research centers in China. Thanks to funding support from MOST, NSFC, etc., the CO_2 storage research group at IRSM, Chinese Academy of Sciences has developed from a very small group with three research staff to the present research center with over 15 large-scale experimental devices and more than 40 full-time research staff, project engineers, and postdoctoral fellows. Now the research focuses of the CO_2 storage research group almost cover all important research fields related to CCUS. The CO_2 storage research group started research on CO_2 corrosion characterization and control in 2010. Early works were focused on scanning electron microscope (SEM) characterization of wellbore cement corrosion induced by high concentration CO_2, identification of cement corrosion products by X-ray diffraction (XRD), etc. Recently, the research areas have been extended to the development of anti-corrosion cement additives, flow-corrosion coupled analysis of the wellbore system, SEM characterization of CO_2-induced wellbore casing corrosion, etc.

This book covers research progress made by the CO_2 storage research group at IRSM-CAS, Carbon Neutrality Institute at China University of Mining and Technology, School of New Energy and Materials at Southwest Petroleum University, College of Petroleum Engineering and Unconventional Petroleum Research Institute at China University of Petroleum (Beijing), and other groups working on CO_2-induced corrosion all over the world. Authors of this book are well-known experts in the field of CO_2 corrosion characterization and control. This book will serve as a good reference for design and operation for corrosion risk minimization of CCUS projects. This book will also provide research directions for postdoctoral fellows and Ph.D. students who recently start their research in CO_2 corrosion characterization and control.

This book systematically discusses the operational stages that have high risk of CO_2-induced corrosion in CCUS projects and related measures for corrosion control. Corrosion control techniques that are included in this book (i.e., CO_2-resisting steel, CO_2-resisting wellbore cement, anti-corrosion coating, and corrosion inhibitors) are beneficial for corrosion control research and engineering practices. This book consists of seven chapters, covering a brief introduction to CCUS and CO_2-induced corrosion in CCUS (Chap. 1); corrosion theory and corrosion characterization techniques (Chap. 2); corrosion in CO_2 capture and transportation (Chap. 3); corrosion in CO_2 geological utilization and storage (Chap. 4); corrosion-resistant steel and cement (Chap. 5); anti-corrosion coating (Chap. 6); and corrosion inhibitors (Chap. 7).

This book was primarily organized by Prof. Liwei Zhang and written by Prof. Liwei Zhang (Chaps. 1, 3, 4, 5, 6), Prof. Shijian Lu (Chap. 3), Prof. Xiaowei Cheng (Chap. 5), Prof. Yongcun Feng (Chaps. 4 and 5), Prof. Wei Yan (Chap. 4), Dr. Manguang Gan (Chap. 7), Dr. Kaiyuan Mei (Chap. 5), Mr. Hanwen Wang (Chap. 2), Mr. Chenyang Deng (Chap. 6), Prof. Hejuan Liu (Chap. 1), and Prof. Yan Wang (Chap. 4). Dr. Manguang Gan and Mr. Hanwen Wang completed proofreading and formatting.

This project was primarily supported by the National Science Foundation of China (Grant No. 42172315 and U1967208), and Key R&D Program of Inner Mongolia Province of China (Grant No. 2021ZD0034).

If there are mistakes in this book due to limits in our knowledge and information, any criticisms and suggestions are welcome.

Wuhan, China
February 2023

Liwei Zhang

Contents

Chapter 1
Background

Liwei Zhang and Hejuan Liu

Scientific evidences have shown that the Earth is experiencing an unprecedented rate of climate warming. From 1850 to 2020, the global average surface temperature increased by about 1.2 °C, and that temperature increase was mainly attributed to human activities (Fig. 1.1). After 1977, the global surface temperature has shown a steep increasing trend. The global average annual temperature in the mid-to-late twentieth century was significantly higher than the global average temperature of the entire twentieth century, and the global surface temperature in 2016 was 0.9 °C higher than the twentieth century average [1]. The latest report of the Intergovernmental Panel on Climate Change (IPCC) pointed out that if no measures are taken, the global average surface temperature will increase by 3.3 ~ 5.7 °C by 2100, compared with the average global surface temperature of the period 1850 ~ 1900 [2].

Global warming has had a significant impact on natural ecosystems and human life on the Earth. The impact is represented in seven aspects: (1) Accelerating the melting of snowfields and glaciers in the North and South Poles, which causes sea level rise and poses a threat to cities and towns in or near coastal areas. (2) Disturbing growth, migration habits, life cycles, etc. of terrestrial and aquatic species. (3) Affecting the water cycle in some areas, and causing extreme precipitation. (4) Leading to higher sea temperatures and lower pH, posing a threat to marine ecosystems and coral reefs. (5) Contributing to the reduction of crop production in some areas. (6) Increasing the frequency of extreme weathers such as droughts and hurricanes. (7) Leading to the softening of the permafrost layer in high latitudes, which poses a threat to the safety of the buildings in high latitudes.

Among the impacts of global warming, the impact of global warming on the North and South Poles should be paid special attention to. Climate models predict that the

---

L. Zhang (✉) · H. Liu
State Key Laboratory of Geomechanics and Geotechnical Engineering, Institute of Rock and Soil Mechanics, Chinese Academy of Sciences, Wuhan 430071, China
e-mail: lwzhang@whrsm.ac.cn

University of Chinese Academy of Sciences, Beijing 100049, China

L. Zhang (ed.), *Corrosion in CO_2 Capture, Transportation, Geological Utilization and Storage*, Engineering Materials, https://doi.org/10.1007/978-981-99-2392-2_1

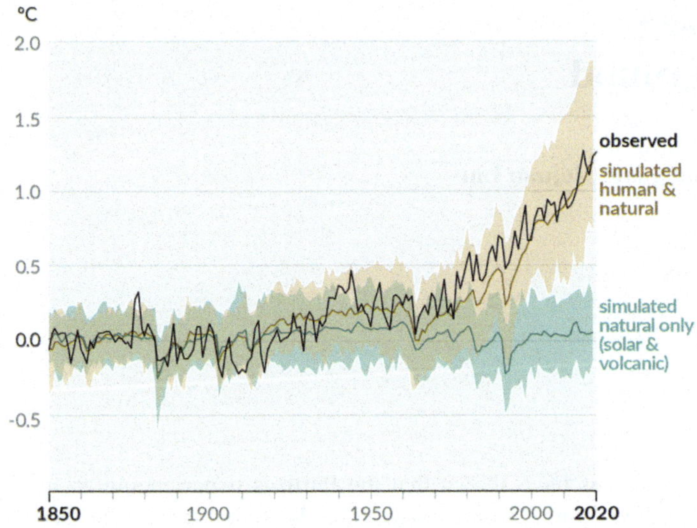

Fig. 1.1 Change in global surface temperature (annual average) as observed and simulated using human and natural and only natural factors (both 1850–2020) [2]

cold polar regions will be affected by global warming in the first place, which is a process called "Arctic acceleration" [3]. The imapct of Arctic acceleration is clear—the Arctic is warming twice as fast as other parts of the world. In less than half a century, from 1971 to 2019, the Arctic's average annual temperature rose by 3.1 °C. This has resulted in the loss of year-round sea ice on the Arctic Ocean. The sea level of the oceans can be raised up to 2 m by the year 2100 due to the melting of the large glaciers and terrestrial ice sheets in polar regions [4]. The oceans could potentially rise by up to 76 m above the sea level at present, if the entire ice volume in Greenland and Antarctica melts [5]. Meteorologists have also pointed out that ocean currents and the paths of storms may also be affected, if large volumes of fresh water enter the oceans from melting ice caps [6].

Atmospheric CO_2 concentration is increasing at an unprecedented rate, which is the most important contributor to global warming. U.S. National Oceanic and Atmospheric Administration (NOAA) and the Scripps Institute of Oceanography jointly built an atmospheric CO_2 observatory on the flank of the Mauna Loa volcano in Hawaii in 1957, with the purpose to record the change of atmospheric CO_2 concentration over time [6]. Nearly two decades of data showed an alarming trend of rising CO_2 levels in the atmosphere [7]. Atmospheric CO_2 concentration has increased by nearly 100 parts per million (ppm) or around 30% over the past 60 years.

Greenhouse gas emissions caused by human activities are the main contributor to global warming. Since the beginning of the First Industrial Revolution in the UK during the mid-18th Century that transitioned the manufacturing of products from individual handcrafting to mass production using machines [8], the use of fossil fuel by humans as an energy resource has been growing. By the early 20th

Century, electricity drove most machines in factories, which also commonly used coal as a power source. A further increase in the use of fossil fuel was driven by the invention of the internal combustion engine in the 19th Century, which required petroleum. Natural gas was adopted in the mid-20th Century as a widespread utility to replace manufactured gas. The increase in atmospheric CO_2 concentrations recorded at Mauna Loa CO_2 observatory fits well with the rise of fossil fuel use since 1957 [6]. This human-produced, or "anthropogenic" CO_2 has raised greenhouse gas levels in the atmosphere and is the primary reason to cause climate change.

In 2019, the content of CO_2 in the Earth's atmosphere reached 410 ppm, and the total amount of all greenhouse gases in the atmosphere reached the highest level in a 800,000-year period [2]. Greenhouse gas refers to the gas in the atmosphere that can absorb the long-wave thermal radiation emitted from the Earth's surface. The Sun's radiation to Earth is mainly short-wave radiation and is not absorbed by greenhouse gases. After absorbing solar short-wave radiation, the Earth radiates heat back into outer space, and the thermal radiation back into outer space from the Earth is mainly long-wave infrared radiation with a wavelength range of 3–30 μm. When such long-wave radiation enters the atmosphere, it is easily absorbed by certain gas molecules, thus preventing the heat from releasing into the outer space. Because these gases prevent heat from being lost from the Earth and play a role similar to the insulation effect of greenhouses, these gases are called greenhouse gases. Major greenhouse gases include CO_2, CH_4, N_2O, HFC-23, HFC-134a, CF_4, SF_6, NF_3, etc. [2]. The most important greenhouse gas is CO_2, which accounts for about 76% of the total greenhouse gas emissions, due to very significant global CO_2 emissions (Fig. 1.2). In 2015, the Paris agreement was developed, which aims to limit global warming to 2 °C by 2100, while attempting to limit the temperature increase to 1.5 °C by cutting emissions of greenhouse gases, especially CO_2 [9].

CO_2 capture, utilization and storage (CCUS) is a process that captures CO_2 emissions from sources like coal-fired power plants and either reuses or stores it so it will not enter the atmosphere [11]. CCUS is one of the most favored approaches to limit the CO_2 emissions and tackle the problem of global warming. CCUS comprises several different processes: separating CO_2 from emission sources, CO_2 transportation, CO_2 conversion and utilization, and underground storage of CO_2 with long-term isolation from the atmosphere [12] (Fig. 1.3).

The separation of CO_2 from emission sources is commonly referred to as CO_2 capture. CO_2 capture from power generation can be divided into three main categories, i.e., pre-combustion CO_2 capture, post-combustion CO_2 capture, and oxy-combustion CO_2 capture [14]. For pre-combustion CO_2 capture, the fuel is partially oxidized and reacted with steam to form a CO_2 and H_2 mixture containing 15–60% CO_2, from which the CO_2 can be separated by liquid absorbent of CO_2 (commonly-used liquid absorbents of CO_2 include amine solution, ionic liquid, etc.) or solid adsorbent of CO_2 [14]. For post-combustion and oxy-combustion CO_2 capture, the fuel is burned with air and oxygen, respectively, and the CO_2 is separated from flue gas. The composition of the flue gas from the post-combustion process is different from that of the oxy-combustion process. In the latter technique, the flue-gas contains highly concentrated CO_2, which requires only minimal further processing before

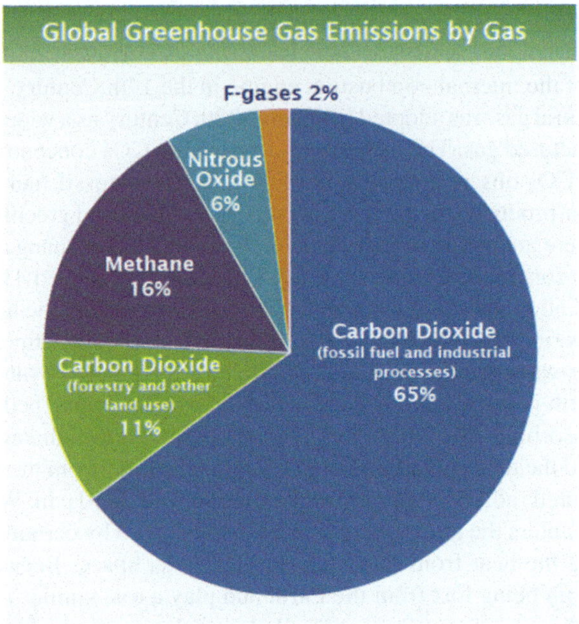

Fig. 1.2 Global greenhouse gas emissions in 2010. Details about the sources included in these estimates can be found in the Contribution of Working Group III to the Fifth Assessment Report of the Intergovernmental Panel on Climate Change [10]

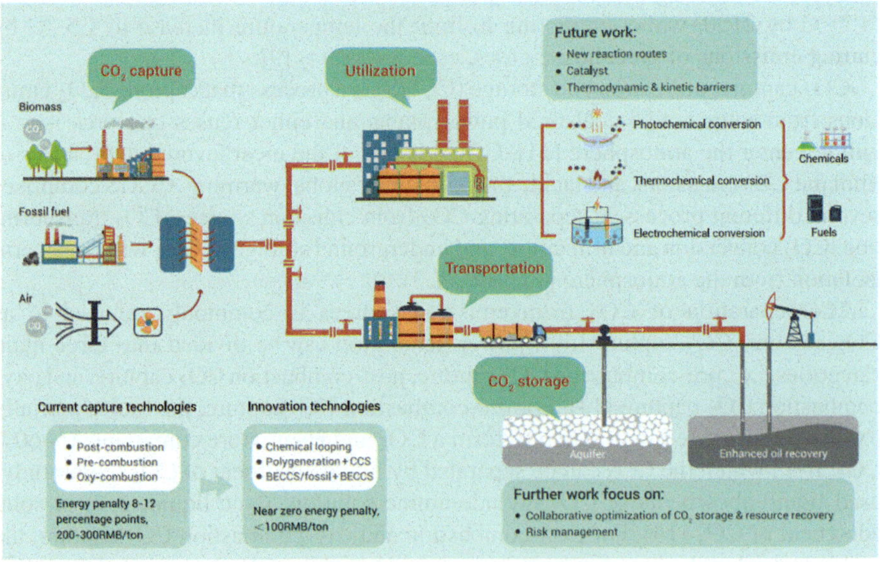

Fig. 1.3 Typical processes in carbon capture, utilization and storage (CCUS) [13]

compressed for transportation and storage, and therefore reduces the cost of CO_2 separation [14].

The CO_2 captured from the emission sources can sometimes be utilized right at the emission sources, but in most cases, the captured CO_2 needs to be transported to suitable sites either for utilization and conversion, or for permanent underground storage. CO_2 from its point of capture needs to be transported by CO_2 transport infrastructure to either an end for CO_2 use or an end for CO_2 storage. Pipelines, ships, trucks, trains and even barges are available for CO_2 transport. Among those CO_2 transport options, pipelines and ships are the most scalable options with the lowest cost per ton of CO_2 [15]. Pipelines that transport CO_2 captured at multiple sources can form a pipeline transportation network that covers a large area, which can support the widespread deployment of CCUS projects. One example of large-scale pipeline CO_2 transport project is the Alberta Carbon Trunk Line in Canada. The Alberta Carbon Trunk Line has a total length of 240 km and has a capacity to transport 14.6 Mt CO_2 per year. Commissioned in 2020, the pipeline had significant excess capacity to allow connection of other facilities to the pipeline in the future. The Alberta Carbon Trunk Line is a typical example of global deployment of multi-user CO_2 pipeline networks. Other examples include the Midwest Carbon Express in the United States, Fluxys-Equinor offshore CO_2 pipeline connecting Belgium with Norway, and the Delta Corridor connecting parts of Germany and the Netherlands. For ships, currently two large-scale CO_2 shipping routes are under construction as part of the Northern Lights project and more are being designed by other projects. On inland waterways, Barges are ideal for CO_2 transport, and some site operators are evaluating the option of CO_2 transport using barges on inland waterways. CO_2 transport by barge or ship requires different CO_2 phase, temperature and pressure conditions, compared with pipeline CO_2 transport. CO_2 terminals are required for waterborne transport in order to load and unload the CO_2 and ensure that it is properly conditioned for further transport and injection. Several CO_2 terminals are in development, including the Port of Antwerp in Belgium, the Port of Gdansk in Poland, the Port of Gothenburg in Sweden, etc. [15].

In a whole-chain CCUS system, all CO_2 captured from emission sources needs to be handled either by CO_2 conversion and utilization, or by underground storage of CO_2. At the current stage, the amount of CO_2 that can be converted by the state-of-the-art CO_2 conversion and utilization technologies like hydrogenation of CO_2 to produce syngas, photocatalytic reduction of CO_2 to fuels, electrochemical CO_2 reduction to fuels, etc. [13] is limited. To deeply cut CO_2 emissions, large amounts of CO_2 need to be captured and stored in deep subsurface formations including saline aquifers, gas and oil reservoirs, coal beds, etc., which is referred to as CO_2 geological utilization and storage (CGUS). CGUS has been used in the United States for more than 50 years, mainly in the form of CO_2-enhanced crude oil recovery. Therefore, CGUS is not a new concept, but is an operable and safe technology proven by engineering practices. For point sources with large amounts of CO_2 emissions (power plants, cement plants, oil refinery plants, etc.) and with ideal geological sinks nearby, underground storage of CO_2 can be regarded as the most effective approach for deep CO_2 emission cut.

In summary, CCUS can achieve near-zero CO_2 emissions from fossil energy consumption, which is of great significance to global greenhouse gas emission reduction. IPCC noted in its 2014 Fifth Assessment Report that the target of limiting atmospheric CO_2 equivalent concentrations to 450 ppm by 2100 will not be achieved if the major CO_2 emitters fail to deploy CCUS to cut CO_2 emissions from fossil energy consumption [10]. Therefore, CCUS will play a vital role in reducing greenhouse gas emissions, and is an indispensable and important part of the measures of all mankind to mitigate global warming and reduce CO_2 emissions. However, success in CCUS requires a globally shared political will and a sense of urgency to create the enabling conditions for the deployment of CCUS on a massive scale [16].

CO_2-induced corrosion may impair the performance of a CCUS system, and characterization and control of CO_2-induced corrosion plays a very important role in safe deployment of CCUS. CO_2 is an acid gas, which forms carbonic acid and causes pH drop when dissolved in water. As a result, a co-existence of CO_2 and water in the environment causes corrosion of various components in the CCUS operation chain, including CO_2 absorption towers, compressors, pipelines, valves, wellbores, monitoring devices, etc. Therefore, it is very important to characterize, monitor and mitigate CO_2-induced corrosion at all stages of the CCUS operation chain. The rate of CO_2-induced corrosion in CCUS scenarios is dependent on many factors, but even a slow corrosion can result in pitting of steel surface, and eventually lead to pinhole leaks. Severe corrosion may cause CO_2 blow-out and lead to project failure. In many cases, the mechanical failure of pipelines, steel casings, cement rings, etc. in CCUS projects is attributed to corrosion. For example, brittle fractures occur where stresses are concentrated as a result of corrosion, causing small cracks that can then grow and form large cracks. CO_2 pipelines are at greater risk of failure from long-running axial brittle fractures as a result of the severe Joule–Thomson cooling and embrittlement that occur around any leaking points caused by corrosion [16]. Due to the complex nature and frequent occurrence of corrosion in CCUS projects, there is an urgent need to systematically discusses CO_2 corrosion and control at all stages of the CCUS operation chain, and that is the motivation for this book.

This book has a comprehensive coverage of the operational stages that have a high risk of CO_2-induced corrosion in whole-chain CCUS projects, and this book has an in-detail discussion on the related measures for corrosion control. Experimental investigations of CO_2 corrosion rates, the impact of impurities on corrosion, development of CO_2 corrosion level prediction models, identification of suitable corrosion inhibitors, etc. have been covered in this book. Some corrosion control techniques that are included in this book (e.g., CO_2-resisting wellbore cement additives, CO_2-resisting steel coating, etc.) are beneficial for corrosion control research and engineering practices.

References

1. NOAA: Climate at a glance: global time series, Silver Spring: National Centers for Environmental information (2018)
2. IPCC: Climate Change 2021: The Physical Science Basis. Contribution of working group I to the sixth assessment report of the intergovernmental panel on climate change, 2391pp. IPCC, Cambridge (2021)
3. IPCC: Climate Change 2013: The Physical Science Basis. Contribution of working group I to the fifth assessment report of the intergovernmental panel on climate change, 1535pp. IPCC, Cambridge (2013)
4. Bamber, J.L., Oppenheimer, M., Kopp, R.E., Aspinall, W.P., Cooke, R.M.: Ice sheet contributions to future sea-level rise from structured expert judgment. Proc. Natl. Acad. Sci. U.S.A. **116**(23), 11195–11200 (2019)
5. Poore, R., Williams J.R.S., Tracey, C.: Sea level and climate: US Geological Survey Fact Sheet 002–00, 2p (2000). https://pubs.usgs.gov/fs/fs2-00/
6. Soeder, D.J.: Greenhouse gas sources and mitigation strategies from a geosciences perspective. Adv. Geo-energy Res. **5**(3), 274–285 (2021)
7. Keeling, C.D., Bacastow, R.B., Bainbridge, A.E., Ekdahl, C.A., Guenther, P.R., Waterman, L.S., Chin, J.F.S.: Atmospheric carbon-dioxide variations at Mauna-Loa observatory, Hawaii. Tellus **28**(6), 538–551 (1976)
8. Wrigley, E.A.: Reconsidering the industrial revolution: England and Wales. J. Interdiscip. Math. **49**(1), 9–42 (2018)
9. Osman, A.I., Hefny, M., Abdel, M.M., Elgarahy, A.M., Rooney, D.W.: Recent advances in carbon capture storage and utilisation technologies: a review. Environ. Chem. Lett. **19**(2), 797–849 (2021)
10. IPCC: Climate Change 2014: Synthesis Report. Contribution of working groups I, II and III to the fifth assessment report of the intergovernmental panel on climate change, 151pp. IPCC, Geneva (2014)
11. U.S. Department of Energy: Carbon capture, utilization & storage. https://www.energy.gov/carbon-capture-utilization-storage
12. Tapia, J.F.D., Lee, J.Y., Ooi, R.E., Foo, D.C., Tan, R.R.: A review of optimization and decision-making models for the planning of CO_2 capture, utilization and storage (CCUS) systems. Sustain. Prod. Consum. **13**, 1–15 (2018)
13. Wang, F., Harindintwali, J.D., Yuan, Z., Wang, M., Wang, F., Li, S., Yin, Z., Huang, L., Fu, Y., Li, L.: Technologies and perspectives for achieving carbon neutrality. The Innovation **2**(4), 100180 (2021)
14. Aghaie, M., Rezaei, N., Zendehboudi, S.: A systematic review on CO_2 capture with ionic liquids: current status and future prospects. Renew. Sust. Energ. Rev. **96**, 502–525 (2018)
15. Rachael, M., Carl.G.: CO_2 transport and storage: infrastructure deep dive. IEA, Paris (2022)
16. Rackley, S.A.: Carbon Capture and Storage. Butterworth-Heinemann, Oxford (2017)

Chapter 2
Corrosion Theory and Corrosion Characterization Techniques

Hanwen Wang

2.1 Corrosion Theory

Corrosion is the deterioration of materials caused by the interaction between materials and the environment, not only the chemical reaction of materials, but also the deterioration of surface properties or mechanical properties [1]. Broadly speaking, any structural material (both metallic and non-metallic) is at risk of corrosion. For example, the corrosion of concrete, the weathering of masonry, the aging of plastics and rubber, and the decay of wood can be included in the category of corrosion. In engineering projects, metal corrosion and cement corrosion are two more common corrosion phenomena, and scholars have studied both more.

2.1.1 Metal Corrosion Theory

During the service process, metal materials usually undergo chemical, electrochemical and physical interactions with the environmental medium in which they are located, resulting in deterioration and destruction, a process called metal corrosion. Since the eighteenth century, scientists have carried out corresponding research work on metal corrosion and after nearly 200 years of development, a complete theoretical framework of corrosion has been formed.

H. Wang (✉)
State Key Laboratory of Geomechanics and Geotechnical Engineering, Institute of Rock and Soil Mechanics, Chinese Academy of Sciences, Wuhan 430071, China
e-mail: churchwang@163.com

University of Chinese Academy of Sciences, Beijing 100049, China

L. Zhang (ed.), *Corrosion in CO_2 Capture, Transportation, Geological Utilization and Storage*, Engineering Materials, https://doi.org/10.1007/978-981-99-2392-2_2

2.1.1.1 Thermodynamics

From a thermodynamic point of view, in addition to a few precious metals such as Au, Pt, in the natural environment and many corrosive media, the vast majority of metals have thermodynamic instability. Under certain conditions, these metals will undergo a transition process from an atomic state to an ionic state, that is, a corrosion process. Metal corrosion processes are generally carried out under open system conditions at constant temperature and pressure. According to the principle of thermodynamics, the Gibbs free energy criterion can be used to determine the direction and limit of corrosion reaction, that is, under isothermal and isobaric conditions.

$$\begin{aligned} \Delta G_{T,p} &< 0 \quad \text{spontaneous process} \\ \Delta G_{T,p} &= 0 \quad \text{state of equilibrium} \\ \Delta G_{T,p} &> 0 \quad \text{non-spontaneous process} \end{aligned} \tag{2.1}$$

where G is the Gibbs free energy. Under the condition of constant temperature and pressure, the change of corrosion reaction free energy can be calculated by the chemical sites of reactants and products involved in the reaction, such as corrosion reaction:

$$\mathrm{M_{(s)}} + \mathrm{O} \rightarrow \mathrm{M^{n+}_{(aq)}} + \mathrm{R} \tag{2.2}$$

The free energy change of the reaction is:

$$\Delta G_{T,p} = \sum \nu_i \mu_i = \nu_{M^{n+}} \mu_{M^{n+}} + \nu_R \mu_R - \nu_M \mu_M - \nu_O \mu_O \tag{2.3}$$

where ν_i is the stoichiometric number of the substance i in the corrosion reaction formula, and the μ_i is the chemical potential of the ith substance. For corrosion reactions in solution, μ_i can be calculated from the following equation:

$$\mu_i = \mu_i^\theta + 2.3RT\lg a_i = \mu_i^\theta + 2.3RT\lg \gamma_i C_i \tag{2.4}$$

Where a_i, γ_i, and C_i are the activity, activity coefficient and concentration of the ith substance, respectively; R is the gas constant, T is the absolute temperature, and μ_i^θ is the standardized chemical potential of the ith substance.

From the electrochemical mechanism, the tendency of metal corrosion can be identified by the electromotive force of the corrosion cell which is the main reaction in the corrosion process. Under the condition of constant temperature and pressure, the reaction electromotive force or potential and free energy can be converted to each other:

$$\Delta G_{T,p} = -nFE \tag{2.5}$$

where n is the number or chemical valence of electrons participating in the reaction, F is the Faraday constant (96,500 C/mol), and E is the electromotive force (V). The magnitude of the electromotive force E is determined by the anodic dissolution potential (E_A) and cathodic reduction potential (E_C):

$$E = E_C - E_A \tag{2.6}$$

The larger E of the reaction, the greater the spontaneous tendency of the reaction. According to Eqs. (2.1) and (2.5), the electrochemical criterion of metal corrosion can be inferred as follows:

$$\begin{array}{ll} E_A < E_C & \text{metals with potential } E_A \text{ corrode spontaneously} \\ E_A = E_C & \text{state of equilibrium} \\ E_A > E_C & \text{metals with potential } E_A \text{ do not corrode spontaneously} \end{array} \tag{2.7}$$

It should be noted that the corrosion tendency of metal in aqueous solution is closely related to the activity and concentration of medium in aqueous solution when electrochemical criterion is used. The electromotive force of the corrosion reaction is equal to the algebraic sum of the potential of the anode reaction (metal corrosion reaction) and the potential of the cathode reaction. Take the corrosion reaction that can occur in Zn in acid solutions as an example:

Anodic reaction

$$Zn \rightarrow Zn^{2+} + 2e^- \tag{2.8}$$

$$E_{Zn^{2+}/Zn} = E^{\theta}_{Zn^{2+}/Zn} + \frac{RT}{2F}\ln\frac{a_{Zn^{2+}}}{a_{Zn}} \tag{2.9}$$

Cathodic reaction

$$2H^+ + 2e^- \rightarrow H_2 \tag{2.10}$$

$$E_{H^+/H_2} = E^{\theta}_{H^+/H_2} + \frac{RT}{2F}\ln\frac{p_{H_2}}{a^2_{H^+}} \tag{2.11}$$

The reaction of the corrosion system is

$$Zn + 2H^+ \rightarrow Zn^{2+} + H_2 \uparrow \tag{2.12}$$

The electromotive force of the corrosion reaction is

$$E = E_{H^+/H_2} + E_{Zn^{2+}/Zn} = \left(E^{\theta}_{Zn^{2+}/Zn} + E^{\theta}_{H^+/H_2}\right) + \frac{RT}{2F}\ln\frac{a_{Zn^{2+}}p_{H_2}}{a_{Zn}\cdot a^2_{H^+}} \tag{2.13}$$

By analogy, if the form of corrosion reaction is as follows:

$$lL + mM + \cdots \rightarrow pP + qQ + \ldots \tag{2.14}$$

According to the thermodynamic principle, the electromotive force of corrosion reaction can be derived as follows:

$$E = E^{\theta} + \frac{RT}{2F}\ln\frac{a_P^p \cdot a_Q^q \ldots}{a_L^l \cdot a_M^m \ldots} \tag{2.15}$$

Equation (2.15) is the Nernst equation, which reflects the relationship between the electromotive force of corrosion reaction and the activity of reactants and products. Obviously, the condition for corrosion reaction to occur spontaneously is that the electromotive force E of corrosion reaction is greater than zero, and the change of activity of reactants and products can change the value of electromotive force E. In other words, the change of concentration of reactants and products may lead to the change of corrosion tendency.

It is worth noting that any corrosive reaction that is thermodynamically unable to proceed spontaneously is unlikely to occur; For those corrosion reactions that $\Delta G < 0$ or $E > 0$, although they have a tendency to spontaneously carry out thermodynamically, this does not mean that these corrosion reactions can necessarily occur or have a high corrosion rate, because the latter belongs to the kinetic category of problems and should be distinguished from thermodynamic problems.

2.1.1.2 Dynamics

1. The polarization

For a $\Delta G < 0$ corrosion system, when corrosion begins, there will be charge exchange on the interface, that is, static current flows into or out of the electrode surface. At this time, the electrode will no longer be in equilibrium, and the measured potential of the electrode will change. Potential changes are always in the direction of preventing equilibrium from breaking or promoting the formation of a new equilibrium, that is, resisting the flow of current. For the corrosion cell, when there is current flow, the cathode potential changes in the negative direction, and the anode potential changes in the positive direction, so that the potential difference of the original cell decreases, and the corrosion current decreases sharply. This phenomenon is called the polarization of the battery [2]. When there is electrostatic flow on the electrode, the phenomenon that the electrode potential deviates from the equilibrium potential is called electrode polarization [2].

The polarization of the electrode occurs because there is resistance at each step of the electrode reaction, such as mass transfer, electron transfer, etc. When one of the steps is subjected to the largest resistance, its speed will be much slower than the other steps, and the kinetic characteristics of the whole electrode reaction will be the same as that of the slowest step, which is called the control step of the electrode reaction process [3]. According to the connection with the control step, polarization can be divided into concentration polarization, electrochemical polarization and resistance polarization:

(1) Concentration polarization: When current flows through the electrode, the formation of dissolved metal ions can be collected into the bulk solution diffusion, the ion concentration is greater than in the bulk solution concentration near the electrode, like the electrodes immersed in a solution of a large concentration, and the equilibrium electrode potential is usually refers to the corresponding to the concentration of a solution of ontology, the electrode potential is obviously higher than that of the equilibrium potential, This phenomenon is called concentration polarization.
(2) Electrochemical polarization: The anodic process is the transfer of metal ions from the metal matrix to the solution and the formation of hydrated ions. This reaction process requires a certain amount of activation energy, so that the dissolution rate of the anode is slower than the transfer rate of electrons. The reaction rate of metal ions entering the solution is less than the transfer rate of electrons from the anode to the cathode through the conductor, and too much positive charge accumulates on the anode, leading to the electrode potential moving in the positive direction. This polarization due to the sluggishness of the electrochemical reaction itself is called electrochemical polarization.
(3) In the process of electrode reaction, when the oxide film or corrosion product film is generated on the metal surface, the metal ion or hydration ion generated by the anodic reaction will have a great resistance through this film. The anode current creates a large voltage drop in the film, which makes the potential significantly positive, and the resulting polarization is called resistive polarization.

In a word, polarization will make the electrode potential become more positive or negative, hinder the progress of metal corrosion, reduce the rate of metal corrosion, for the prevention of metal corrosion is beneficial. If you want to reduce the polarization of the electrode, you must provide the electrolyte solution with substances that are easy to react on the electrode, so that the polarization on the electrode is limited to a certain extent. This effect is called depolarization effect, and the substances that reduce the polarization are called depolarizers.

2. Cathodic process of corrosion

The fundamental reason for electrochemical corrosion of metals is that the solution contains substances that can oxidize metals, namely depolarizers, which form an unstable corrosion cell system with metals. The anodic dissolution process of metals is always accompanied by the cathodic process. If there is no corresponding cathodic reaction, the metal will not be corroded. In many cases, the cathodic reaction process plays a decisive role in the corrosion rate. Theoretically, as long as the reduction reaction can absorb the electrons in the metal, it can become the cathodic process of metal electrochemical corrosion [4]. Among all the reactions, hydrogen ion reduction (hydrogen evolution corrosion) and oxygen molecule reduction (oxygen absorption corrosion) are the two most common cathodic depolarization processes.

The reaction mechanism of hydrogen evolution corrosion has been fully studied. In acidic solution, hydrogen evolution reaction steps on metal surface are as follows:

(1) The hydrogen ion H_3O^+ diffuses from the bulk solution to the vicinity of the electrode and transfers to the electrode.
(2) Hydrogen ions H_3O^+ discharge on the electrode.

   a. after discharging on the metal electrode surface, H_3O^+ becomes hydrogen atom adsorbed on the electrode surface H_{ad} (Volmer reaction).

$$H^+ + e^- \rightarrow H_{ad} \tag{2.16}$$

   b. H_3O^+ reacts with hydrogen atoms adsorbed on the electrode surface to form hydrogen molecules (Heyrowsky reaction).

$$H^+ + H_{ad} + e^- \rightarrow H_2 \tag{2.17}$$

(3) The hydrogen atoms adsorbed on the surface of the metal electrode combine into hydrogen molecules (Tafel reaction).

$$2H_{ad} \rightarrow H_2 \tag{2.18}$$

(4) Hydrogen molecules diffuse into the solution from the electrode or escape from the electrode surface by forming bubbles.

In the alkaline solution, water molecules are reduced on the surface of the metal electrode. The hydrogen evolution reaction on the surface of the electrode is carried out as follows:

(1) Water molecules arrive at the electrode and hydroxide ions leave the electrode.
(2) Water molecules are ionized and hydrogen ions are reduced to form hydrogen atoms adsorbed on the metal surface.

$$H_2O \rightarrow H^+ + OH^- \tag{2.19}$$

$$H^+ + e^- \rightarrow H_{ad} \tag{2.20}$$

(3) Compound desorption or electrochemical desorption of adsorbed hydrogen atoms.

$$H_{ad} + H_{ad} \rightarrow H_2 \tag{2.21}$$

$$H^+ + H_{ad} + e^- \rightarrow H_2 \tag{2.22}$$

(4) Hydrogen molecules form bubbles and escape from the electrode surface.

On some metal electrodes, part of the adsorbed hydrogen atoms may diffuse into the interior of the metal, resulting in hydrogen embrittlement during the corrosion process.

The cathodic process of the reduction of oxygen molecules on the cathode is called oxygen absorption corrosion. In contrast to hydrogen ion reduction, oxygen reduction can occur at a corrected electrode potential. Therefore, oxygen corrosion is more common than hydrogen evolution corrosion. Oxygen absorption corrosion can be roughly divided into two processes: oxygen transport and ionization:

(1) Oxygen transport process

a. oxygen enters the solution from the air and transfers to the solution diffusion layer on the metal surface
b. oxygen passes through the solution diffusion layer on the metal surface and reaches the metal surface to form adsorbed oxygen

(2) Oxygen ionization process

The steps of oxygen absorption corrosion reaction are related to pH conditions. In neutral or alkaline solution, the intermediate product of the reaction is hydrogen peroxide ion HO_2^-, and the reaction steps are as follows:

a. forming half-valent oxygen ion O_2^-

$$O_2 + e^- \rightarrow O_2^- \quad (2.23)$$

b. forming hydrogen peroxide ion HO_2^-

$$O_2^- + H_2O + e^- \rightarrow HO_2^- + OH^- \quad (2.24)$$

c. forming hydroxyl ion OH^-

$$HO_2^- + H_2O + 2e^- \rightarrow 3OH^- \quad (2.25)$$

In acidic solution, the intermediate products of the reaction are hydrogen peroxide ion HO_2^- and hydrogen peroxide H_2O_2. The reaction steps are as follows:

a. forming half-valent oxygen ion O_2^-

$$O_2 + e^- \rightarrow O_2^- \quad (2.26)$$

b. forming hydrogen dioxide HO_2

$$O_2^- + H^+ \rightarrow HO_2 \quad (2.27)$$

c. forming hydrogen peroxide ion HO_2^-

$$HO_2 + e^- \rightarrow HO_2^- \tag{2.28}$$

d. forming hydrogen peroxide H_2O_2

$$HO_2^- + H^+ \rightarrow H_2O_2 \tag{2.29}$$

e. forming water H_2O

$$H_2O_2 + 2H^+ + 2e^- \rightarrow 2H_2O \tag{2.30}$$

For most metals, the oxygen reduction reaction on the surface is carried out as described above. However, on some metal oxides, oxygen reduction reaction does not produce hydrogen peroxide or hydrogen peroxide ion and other intermediates, but adsorbed oxygen or surface oxide as intermediates. Since the mechanism of this reaction is not fully studied and some details are still controversial, it will not be described in this chapter.

3. Anodic process of corrosion

In the process of corrosion, the anodic dissolution reaction of metal can be simply expressed as:

$$X \rightarrow X_{sol}^{n+} + ne^- \tag{2.31}$$

However, in fact, the anodic dissolution of metal is a very complex process. On the one hand, the reaction is related to the surface state, structure and performance of metal, on the other hand, it is also affected by the corrosion medium, flow state and other factors [5]. For metal materials, in general, in addition to grains and grain boundaries, there are also various defects on the surface of the metal, such as dislocations, etc. When the metal is subjected to external stress or deformation, these defects provide conditions for metal atoms to leave the lattice and ionize. If there are components in the corrosive medium that can be adsorbed, then the adsorption will also affect the ion process of the metal. When metal ions do not form a precipitate, the medium is flowing and the formed metal ions will leave the solution near the metal surface and enter the bulk solution. At this time, if the anode dissolves quickly, the concentration of metal ions in the solution on the metal surface will increase rapidly, which may form a precipitate film or a special viscous layer. When metal ions form precipitation, they will cover the metal surface, making the metal surface form a non-uniform condition, and at the same time, the cathodic reaction on the precipitation will also change, resulting in more complex corrosion process.

4. Passivation

The phenomenon that the chemical stability of metal or alloy is significantly enhanced by some factors is called metal passivation [6]. There are two kinds of metal passivation, one is chemical passivation, the other is electrochemical passivation. Chemical passivation refers to the passivation phenomenon caused by the chemical interaction between metal and some passivation agents (oxidants), such as concentrated HNO_3, concentrated H_2SO_4, $KMnO_4$, O_2, etc., also known as self-passivation. Electrochemical passivation refers to the phenomenon that the potential of metal is raised to a certain value (i.e. anodic polarization) by applied anode current, so that the metal is changed from activated state to passivation state. In fact, there is no essential difference between chemical passivation and electrochemical passivation, but different measures are adopted to polarize the potential of metal electrodes. The passivated metals all have common characteristics:

(1) When the metal is in the passivation state, its corrosion rate is very low, which can prevent the metal from being corroded. When the metal is transformed from the activated state to the passivation state, a thin film with semiconductor properties can be formed on the surface of the metal, which can inhibit metal dissolution. This thin film is called passivation film. Although the passivation film is very thin, the thickness is generally between 1 and 10 nm, but its corrosion resistance in the medium is good, can keep the anodic dissolution rate of metal at a low value.
(2) The enhanced corrosion resistance of metal after passivation is due to the change of metal surface state, which is an interface phenomenon of metal. When the passivation occurs, the thermodynamic properties of the metal do not change, but the anodic reaction process on the metal surface is greatly hindered. Therefore, the passivation of the metal is a kinetic behavior rather than a thermodynamic behavior.

2.1.2 Cement Corrosion Theory

After hardening, cement can still gradually harden and increase its strength under normal service conditions. But in some corrosive media, the structure of cement will be destroyed, its strength and durability are reduced, or even completely destroyed, this phenomenon is called cement corrosion [7]. Cement corrosion is generally divided into 4 types: soft water corrosion, salt corrosion, acid corrosion and strong alkali corrosion.

2.1.2.1 Soft Water Corrosion

Soft water refers to water containing no or only a small amount of bicarbonate, such as rainwater, distilled water, condensed water and part of river water, lake water, etc. When the cement a long-term contact with soft water, calcium hydroxide by dissolution and erosion of cement, leading to concrete porosity increases, when the cement of free $Ca(OH)_2$ concentration is reduced to a certain extent, other calcium minerals in the cement may also decomposition and dissolution, resulting in the decrease of strength of concrete structure, this kind of phenomenon is called the soft water corrosion of cement.

Among all kinds of hydration products, $Ca(OH)_2$ dissolves the most, so it dissolves first, which not only increases the porosity of cement and makes water easier to penetrate, but also causes the decomposition of hydration products in turn due to the reduction of $Ca(OH)_2$ concentration, such as highly alkaline calcium silicate hydrate and calcium aluminate hydrate into low-alkaline hydration products. And eventually into silicic acid gel, aluminum hydroxide and other materials without gelation ability. In the case of static water and non-pressure water, because the surrounding soft water is easy to dissolve calcium hydroxide saturation, so that the dissolution effect is stopped, so the cement has little effect. However, under the action of flowing water and pressure water, the dissolution of hydration products will continue, and the destruction of cement structure will continue from the surface to the inside.

2.1.2.2 Salt Corrosion

Sulfate with $Ca(OH)_2$ reaction of cement plaster, this kind of product reacts with the hydrated calcium aluminate cement produce calcium rocky stone, its size is about 2.5 times the size of the original hydrated calcium aluminate, so that the solid phase of hardening cement volume increased a lot, the crystallization of a considerable pressure, causing cracking of cement or even destroyed.

$$4CaO \cdot Al_2O_3 \cdot 12H_2O + 3CaSO_4 + 20H_2O \rightarrow 3CaO \cdot Al_2O_3 \cdot 3CaSO_4 \cdot 31H_2O + Ca(OH)_2 \tag{2.32}$$

Magnesium salts generally exist in the form of MgC_2 and $MgSO_4$, and these magnesium salts react with $Ca(OH)_2$ in cement to produce soft $Mg(OH)_2$ without gelling ability.

$$MgCl_2 + Ca(OH)_2 \rightarrow Mg(OH)_2 + CaCl_2 \tag{2.33}$$

$$MgSO_4 + Ca(OH)_2 + 2H_2O \rightarrow Mg(OH)_2 + CaSO_4 \cdot 2H_2O \tag{2.34}$$

2.1.2.3 Acid Corrosion

For general acids, the dissociated H^+ and acid ion (M^-) react with $Ca(OH)_2$ in cement to form H_2O and calcium salts, respectively.

$$Ca(OH)_2 + 2H^+ + 2M^- \rightarrow 2H_2O + CaM_2 \tag{2.35}$$

For carbonic acid, carbonic acid reacts with $Ca(OH)_2$ in cement first, and after neutralization, the cement is carbonized to generate $CaCO_3$. $CaCO_3$ then reacts with carbonic acid to generate $Ca(HCO_3)_2$, which is easily soluble in water. $Ca(HCO_3)_2$ is lost with water, thus destroying the structure of cement.

$$Ca(OH)_2 + H_2CO_3 \rightarrow 2H_2O + CaCO_3 \tag{2.36}$$

$$CaCO_3 + H_2CO_3 \rightarrow Ca(HCO_3)_2 \tag{2.37}$$

2.1.2.4 Alkali Corrosion

Alkaline solutions are generally harmless in low concentrations. However, Portland cement with higher aluminate content will also be destroyed by strong alkali (such as NaOH). NaOH interacts with the unhydrated aluminate in cement clinker to produce soluble sodium aluminate:

$$3CaO \cdot Al_2O_3 + 6NaOH \rightarrow 3Na_2O \cdot Al_2O_3 + 3Ca(OH)_2 \tag{2.38}$$

When the cement is soaked by NaOH and dried in the air, it interacts with CO_2 in the air to produce sodium carbonate:

$$2NaOH + CO_2 \rightarrow Na_2CO_3 + H_2O \tag{2.39}$$

Sodium carbonate crystallizes and deposits in the cement capillary pores, causing the cement to crack.

2.2 Basic Principles of Corrosion Characterization

2.2.1 Principles of Metal Corrosion Characterization

After corroded, metal's weight, surface morphology, etc. will change accordingly, and the corrosion status of the metal can be obtained by characterizing the metal from the above aspects.

Metal corrosion process can be simply summarized as the dissolution of the metal and corrosion products generated, during which the weight of the metal will change accordingly, the metal surface corrosion products will be removed and weighed, the weight change before and after corrosion is obtained, and finally the corrosion rate of the metal during the corrosion cycle is calculated. However, this characterization method can only reflect the average corrosion rate of the metal during the corrosion cycle, and it is impossible to know the morphology of the corrosion products of the metal during the corrosion process and its influence on the corrosion process. With the development of microscopy technology, this problem has been solved. In order to better observe the morphology and structure of corrosion products on metal surfaces, scanning electron microscopy (SEM) or transmission electron microscopy (TEM) can usually be used to enlarge corrosion products by thousands to tens of thousands of times to observe. In addition to surface topography observation, the analysis of the chemical elements and material composition of corrosion products is also an indispensable part of characterizing the metal corrosion process.

2.2.2 Cement Corrosion Characterization Principle

After corroded, cement's microstructure and macroscopic mechanical properties will change, such as corrosion depth, microstructure and phase, average pore size, permeability, compressive strength, etc., which can characterize cement corrosion from the above aspects.

The characterization of cement corrosion is similar to that of metal corrosion, which also observes its corrosion morphology and chemically analyzes the corrosion products. However, since cement has a porous structure, its internal structural characteristics are the focus of characterization and analysis. The internal microstructure of cement can be observed through electron microscopy, but it is limited to a certain plane structure, and it is impossible to achieve three-dimensional characterization of the overall structure of cement. CT scanning technology solves this problem by observing and analyzing the three-dimensional stereoscopic images obtained after scanning, so as to characterize the corrosion condition of the cement as a whole.

2.3 Corrosion Characterization Techniques

2.3.1 Metal Corrosion Characterization Technology

2.3.1.1 Weight Method

Gravimetric method is the most basic, effective and reliable quantitative evaluation method to study corrosion resistance of metals. The gravimetric method is divided into weight gain method and weight loss method. Both of them represent the corrosion rate by the weight difference before and after corrosion of the sample, so as to judge the corrosion degree. The weight gain method is to weigh the sample together with all corrosion products after the corrosion test (Eq. 2.40), while the weight loss law is to weigh the sample after removing all corrosion products (Eq. 2.41). When using gravimetric method, factors such as thickness, compactness and whether corrosion products are easy to fall off should be considered before deciding to choose weight gain or weight loss method to characterize metal corrosion. For loose corrosion products, easy to fall off and easy to remove the metal, usually can use the weight loss method; For the metal with dense corrosion products, good adhesion and difficult to remove, the weight gain method can be used.

$$v_{\text{weightloss}} = \frac{W_0 - W_1}{S_0 t}(\text{g/m}^2\text{h}) \tag{2.40}$$

$$v_{\text{weightincrease}} = \frac{W_2 - W_0}{S_0 t}(\text{g/m}^2\text{h}) \tag{2.41}$$

where, W_0, W_1 and W_2 are the original weight of the sample, the weight of the sample after removing the corrosion products and the weight of the sample without removing the corrosion products (g), S_0 is the surface area of the sample (m^2), and t is the corrosion time.

2.3.1.2 Macro Observation

Before and after metal corrosion, the metal can be observed and analyzed by naked eye. The observer shall pay attention to the shape, distribution, thickness, color, compactness and adhesion of corrosion products. In addition to corrosion products, attention should also be paid to the change of corrosion medium, including the color of the solution, the volume of the solution and whether there is precipitation in the solution. Although the results obtained by the naked eye are crude, they are important for subsequent detailed observation and analysis.

2.3.1.3 Microscopic Observation

Microscopic observation refers to metallographic examination or fracture analysis of the corroded metal, or observation and analysis of the microstructure and phase composition of the corrosion products, so as to study the corrosion characteristics and corrosion kinetics of the metal microstructure. Common microscopic observation methods for metal corrosion include Scanning electron microscope (SEM), Transmission electron microscope (TEM), Energy dispersive spectroscopy (EDS) and X-ray Diffraction (XRD), etc.

Scanning electron microscopy (SEM) is a point-by-point grid scanning image of the specimen surface by focusing electron beam driven by coil. The imaging signal is secondary electron, backscattered electron or absorbed electron. Figure 2.1 shows the structural principle of scanning electron microscopy: the secondary electron signal is collected by the detector and converted into a signal. After processing, the secondary electron image of the surface morphology of the reaction sample is obtained [8]. Backscattered electron imaging reflects the elemental distribution of the sample and the contours of the component regions of different phases. In addition, due to the short de Broglie wavelength of electrons, the resolution is much higher than that of light microscopy, which can reach 0.1 ~ 0.2 nm, and the magnification is from tens of thousands to millions of times. Figure 2.2 shows the cross-sectional morphology of N80 steel corroded at different temperatures for 72 h in CO_2 environment. It can be observed that the corrosion product film is composed of a very thin inner dense layer and an outer layer with defects [9].

The electron beam generated by transmission electron microscope (TEM) is uniformly illuminated to the area to be observed on the sample after being converged by the convergent lens. The incident electrons interact with the sample material. Because the sample is very thin, most of the electrons penetrate the sample, and their intensity distribution corresponds to the morphology, organization and structure of the observed sample area. The electron projected from the sample is magnified by the three-stage magnetic lens and projected on the fluorescent screen of the observation

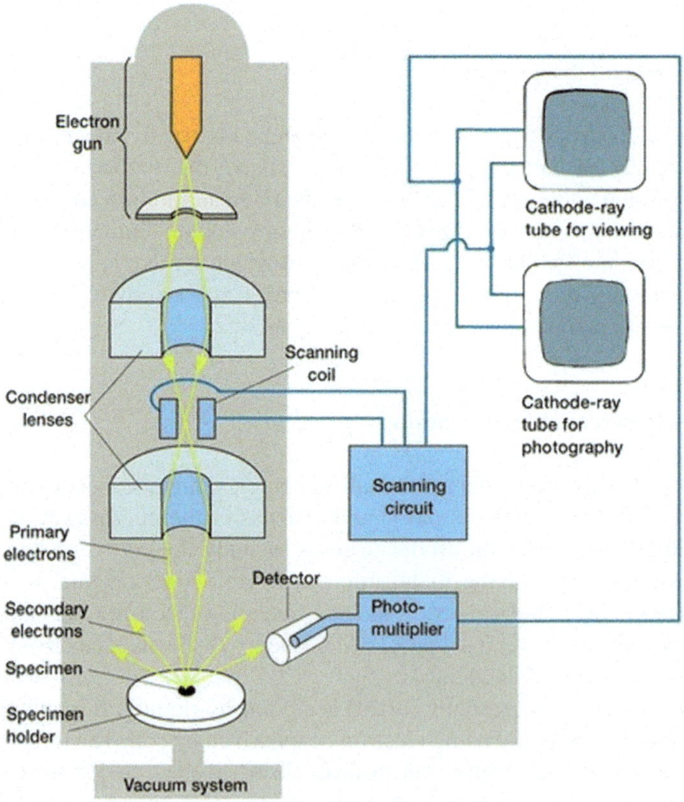

Fig. 2.1 Schematic diagram of SEM structure [8]

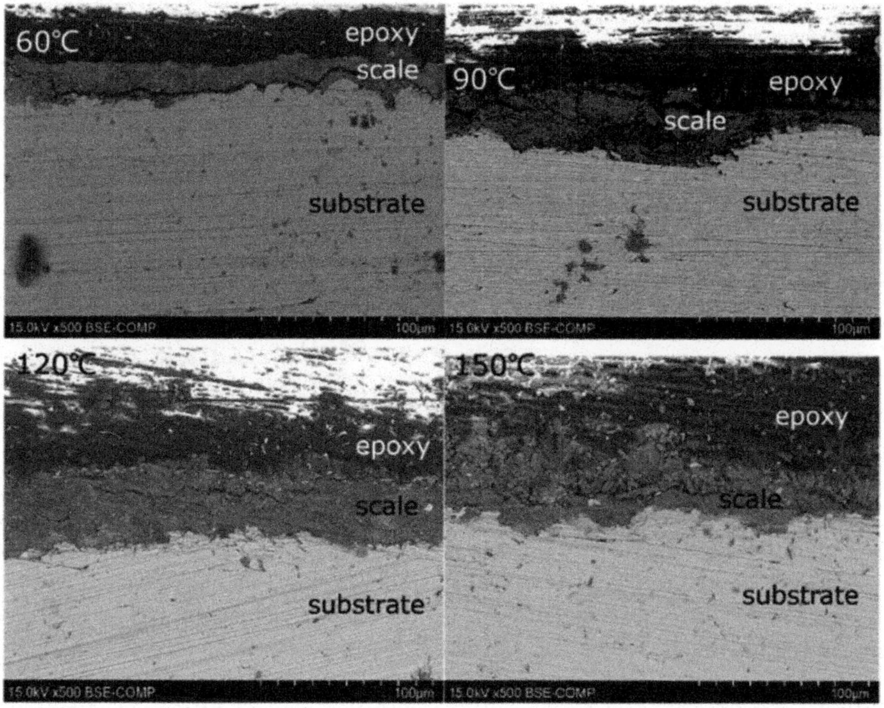

Fig. 2.2 Cross section morphology of N80 steel after 72 h corrosion at different temperatures [9]

figure. The fluorescent screen converts the electron intensity distribution into the light intensity distribution visible to the human eye, so the image corresponding to the morphology, organization and structure of the sample is shown on the fluorescent screen.

Energy dispersive spectroscopy (EDS) is the composition analysis of different elements with different X-ray photon energies, which is usually used in conjunction with scanning electron microscopy and transmission electron microscopy. When the X-ray photon enters the detector, a certain number of electron hole pairs are excited in the Si (Li) crystal. Electron hole pairs are collected by a bias applied to both ends of the crystal and converted into current pulses by a preamplifier. The current pulses are converted into voltage pulses through the main amplifier and enter the multichannel pulse height analyzer, which counts pulses by height classification. In this way, a map of X-ray energy distribution can be traced, which is the principle of EDS test. Figure 2.3 shows EDS analysis of corrosion products of X65 steel under CO_2 and H_2S environment, from which the composition and element content of corrosion products can be known [10].

X-ray diffraction phase analysis (XRD) is a research means by which the composition of a material and the structure or morphology of atoms or molecules inside a material can be obtained by X-ray diffraction analysis of its diffraction pattern. When

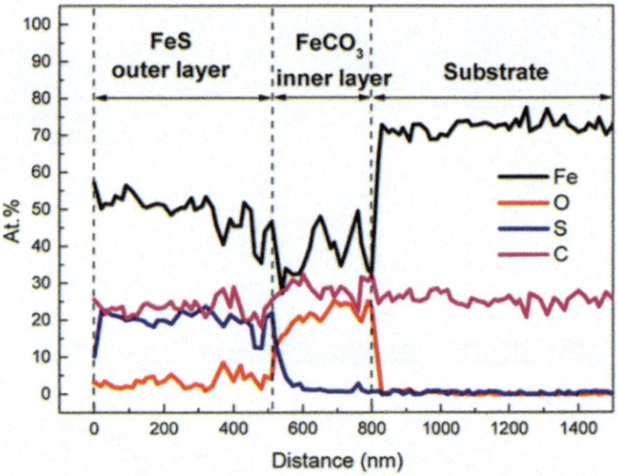

Fig. 2.3 EDS analysis of corrosion products of X65 steel in CO_2 and H_2S environments [10]

an X-ray interacts with a material, its energy conversion is generally divided into three parts: part is scattered, part is absorbed, and part continues to propagate along the original direction through the material. When the wavelength of the scattered X-ray is the same as that of the incident X-ray, diffraction phenomenon will be produced on the crystal, that is, the optical path difference between the crystal plane spacing is equal to an integer multiple of the wavelength. By comparing the characteristic diffraction pattern of each crystal substance with the standard diffraction pattern, the phase present in the sample can be identified by using the three-strong peak principle. XRD can be used for qualitative and quantitative analysis. Qualitative analysis is based on the characteristics of X-ray diffraction spectrum to determine whether the phase exists. Quantitative analysis is based on the intensity distribution of diffraction lines to determine the relative content of each phase. Figure 2.4 shows the XRD analysis of corrosion products of a tubing steel. The main phases of corrosion products can be determined as FeS, $FeCO_3$ and γ-Fe_2O_3 by XRD pattern [11].

2.3.2 Cement Corrosion Characterization Technology

The cement corrosion characterization technology is similar to the metal corrosion characterization technology, which can also be used for macroscopic and microscopic observation. In addition to SEM, EDS, XRD and other characterization technologies mentioned above, cement can also be characterized by spectral analysis, thermal analysis and CT scanning.

Spectral analysis uses light to interact with atoms and molecules of substances to be measured, causing changes in the motion states of molecules in substances, and generating transitions between characteristic energy levels for analysis. When light interacts with material molecules, absorption, emission, scattering and other phenomena can occur. Therefore, the spectrum can be divided into absorption

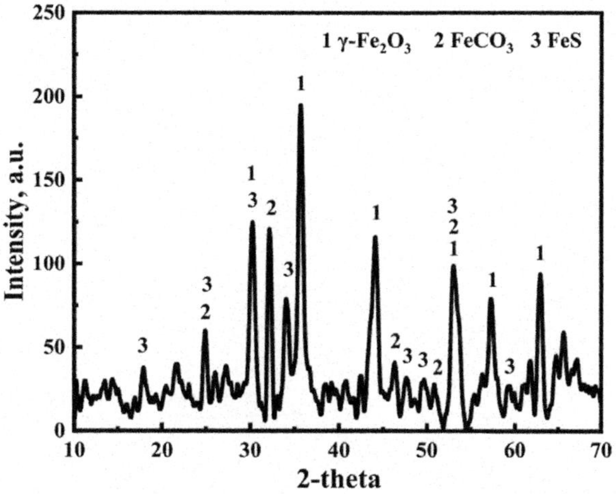

Fig. 2.4 XRD analysis of corrosion products of tubing steel [11]

spectrum, emission spectrum and scattering spectrum. Infrared spectrum analysis and Raman spectrum analysis belong to absorption spectrum and scattering spectrum respectively. Infrared spectrum is to irradiate a beam of infrared rays of different wavelengths to the molecules of a substance, and some infrared rays of specific wavelengths are absorbed to form the infrared absorption spectrum of this molecule. Infrared spectroscopy can be used to study the structure and chemical bonds of molecules, as well as to characterize and identify chemical species. When a monochromatic light with a wavelength much smaller than the particle size of the sample is used to illuminate the gas, liquid or transparent sample, most of the light will be transmitted in the original direction, while a small part will be scattered at different angles, resulting in scattered light. Scattering is the result of the collision between photons and molecules. Raman spectroscopy is the study of the frequency of scattered light produced by the interaction between molecules and light. Similar to infrared spectroscopy, Raman spectroscopy can also be applied to the identification of substances and the analysis of molecular structure. Figure 2.5 is the Fourier infrared spectrum of cement before and after carbonization [12], and Fig. 2.6 is the Raman spectrum of Portland cement [13].

Thermal analysis is a technique for analyzing material parameters as a function of temperature. The properties, compositions, structures, phase transitions and chemical reactions of materials and systems can be determined by thermal analysis. Among all thermal analysis methods, differential scanning calorimetry (DSC), thermogravimetry (TGA), static mechanical thermo-mechanical method (TMA) and dynamic mechanical thermal analysis (DMA) are widely used. DSC measures the change of energy, TGA measures the change of mass, and TMA measures the change of size. DMA measures changes in mechanical properties. Figure 2.7 shows the TGA of dried cement and hydrated cement [14].

CT scanning is a three-dimensional imaging technology that displays the differences in material structure by measuring the X-ray attenuation coefficients of objects

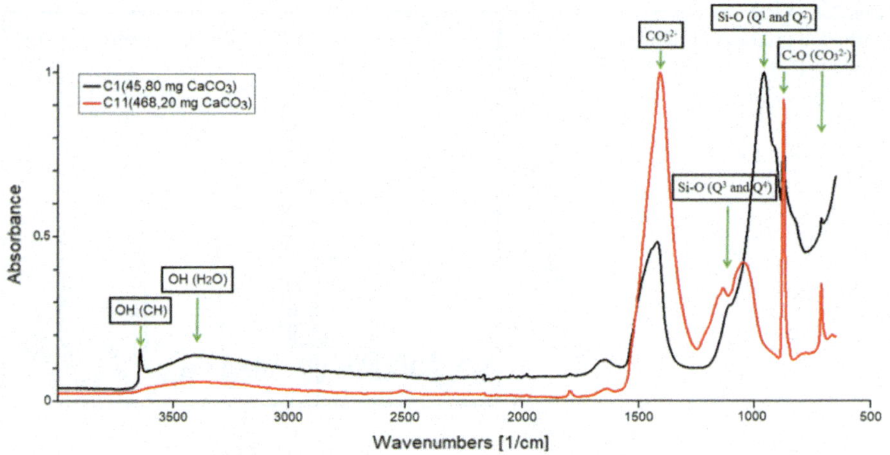

Fig. 2.5 Fourier infrared spectra of cement before (C1) and after (C11) carbonization [12]

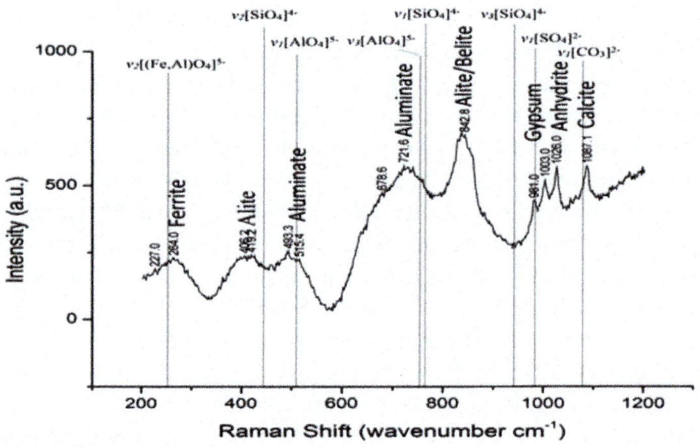

Fig. 2.6 Raman spectra of Portland cement [13]

from different angles and obtaining the spatial distribution information of X-ray attenuation coefficients. CT scanning usually adopts conical X-ray beam and planar array detector to realize projection imaging from different perspectives by rotating the sample. Its working principle is as follows: The conical monochromatic X-ray beam emitted by the X-ray tube irradiates to the test sample, and the transmission intensity of the X-ray is changed after it passes through the material. The detector behind the sample receives the transmitted X-ray, converts it into the projection image signal, and then the projection data is stored on the computer by the image acquisition system. During the scanning process, the samples were rotated 360° around the axis on the workbench according to a certain step distance, so as to obtain

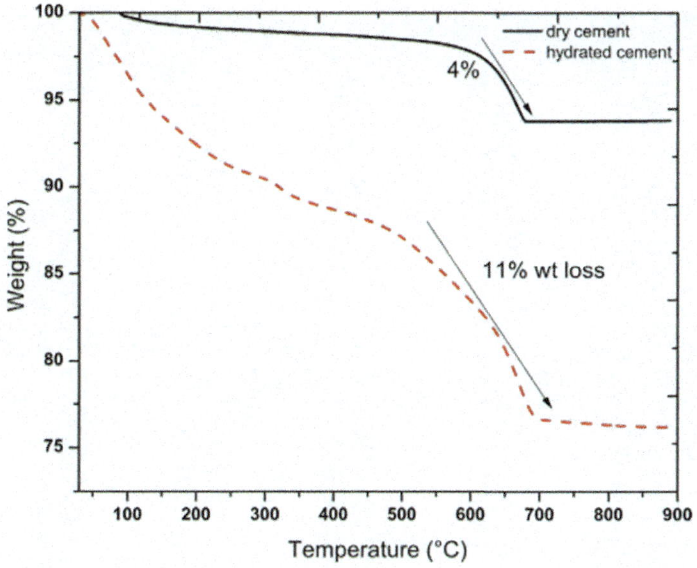

Fig. 2.7 TGA of dried cement and hydrated cement [14]

the projection signals from different perspectives. After the 360° scanning, various image reconstruction algorithms were used to reconstruct the tomography sequence images. Figure 2.8 shows the principle diagram of the CT scanning system [15], and Fig. 2.9 shows the three-dimensional view of the cement sample before and after CO_2 corrosion [16].

CT scanning technology is a non-invasive characterization method and is a powerful tool to provide information about the evolution of porosity of a non-metal porous medium experiencing CO_2 corrosion. There are typically three steps to determine the porosity of a porous medium using CT scanning [17]. First, deionized water is used to clean the sample. After cleaning, the sample is placed in an oven to remove moisture. The sample after drying in the oven is then placed on the objective stage of the CT scanner. During the scanning process, different sections of the sample are

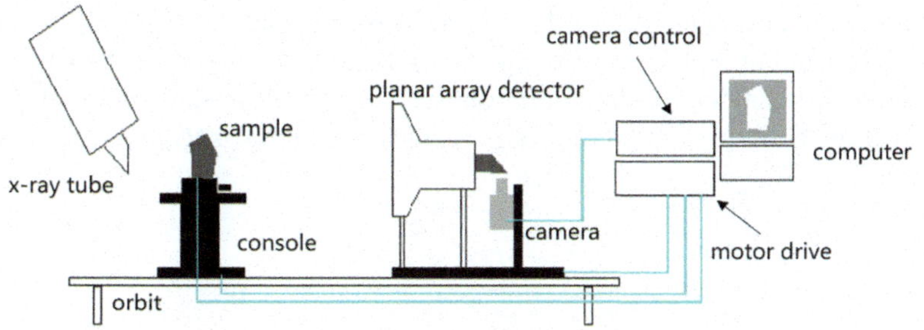

Fig. 2.8 Schematic diagram of the CT scanning system [15]

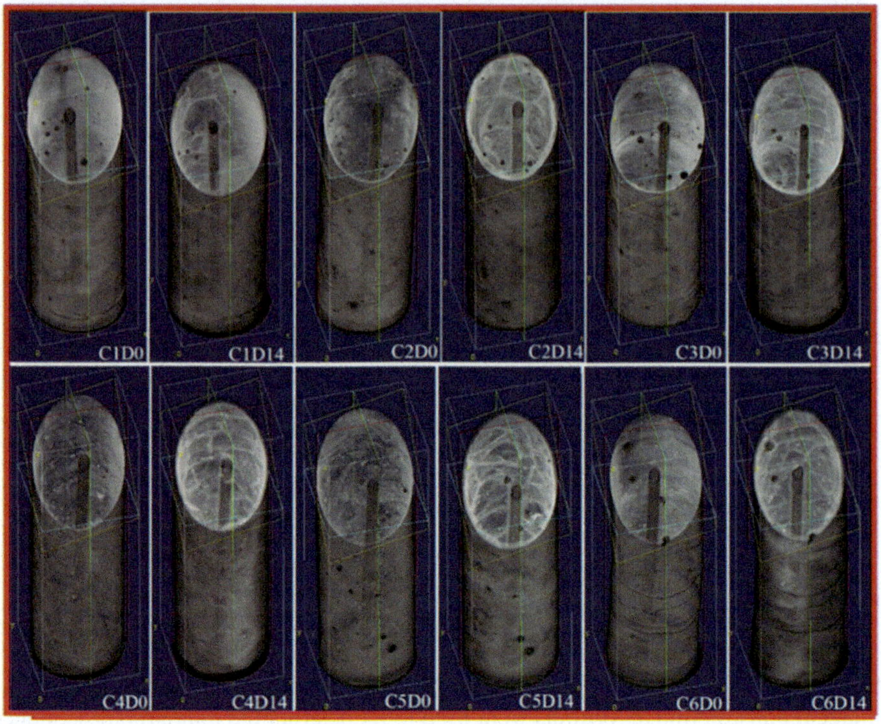

Fig. 2.9 Three-dimensional view of cement before and after CO_2 corrosion [16]. C1 to C6 are sample numbers, "D0" refers to the sample before CO_2 corrosion, and "D14" refers to the sample after 14 days of CO_2 corrosion

scanned, and Eq. (2.42) is applied to calculate the CT number ($CT_{r,d}$) of each voxel for the oven-dried sample. Second, the same sample is placed in a sample container filled with water, and the sample container with the water-saturated sample is fixed on the objective stage of the CT scanner. The scanning parameters and the scanning position of the water-saturated sample need to be the same as those of the dry-sample case. After another round of scanning, the CT number ($CT_{r,w}$) of each voxel of the water-saturated sample is obtained. Third, under the same scanning conditions as the previous two scans, two additional scans are performed to obtain the CT number of water (CT_w) and the CT number of air (CT_a). Finally, the porosity of each voxel of the sample (ϕ) can be calculated according to Eq. (2.43) [18]. The porosity of the entire sample is simply the average of porosities of all the voxels in the CT scanning images.

$$\mathrm{CT} = 1000 \times \frac{\mu - \mu_w}{\mu_w} \tag{2.42}$$

$$\phi = \frac{CT_{r,w} - CT_{r,d}}{CT_w - CT_a} \tag{2.43}$$

CT scanning has been adopted by researchers to study the formation and expansion processes of corrosion products. For example, Fang [19] applied X-ray micro CT to trace the corrosion features of steel bar in cementitious materials (e.g., concrete). Due to a very large difference between the attenuation coefficient of the steel bar and the attenuation coefficient of the steel corrosion products, the formation and expansion of steel corrosion products can be clearly seen (Fig. 2.10). Therefore, X-ray micro CT can track time-dependent development of corrosion products and the subsequent initiation and propagation of corrosion-induced cracks. Xue [20] studied the corrosion process of concrete samples induced by CO_2, with 100 kPa, 500 kPa and 1000 kPa CO_2 partial pressures. It was found that when the CO_2 partial pressure was 0.1 MPa, the volume of pores in 600–1000 μm (macropores) in concrete was reduced after carbonation due to $CaCO_3$ precipitation, and there was no change for the pores with a radius less than 100 μm. Overall, the volume of larger pores was reduced and the compressive strength of concrete was maintained. However, when the CO_2 partial pressure was raised to 1 MPa and there was water presence, the CO_2 carbonation caused excessive accumulation of $CaCO_3$ in the pores, eventually leading to expansion and cracking of pores. The expansion and cracking of pores caused the compressive strength of concrete to decrease [20].

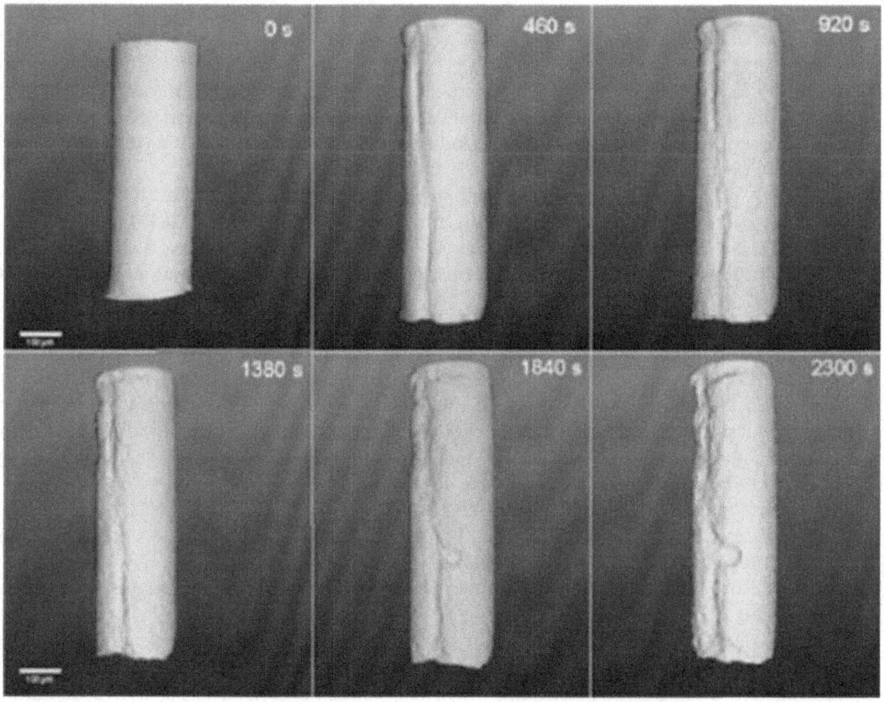

Fig. 2.10 3D micro CT scanning images of the surface of embedded steels in different accelerated corrosion time [19]

References

1. Evans, U.R.: The Corrosion of Metals. Edward Arnold & Company, London (1926)
2. Aikens, D.A.: Electrochemical methods, fundamentals and applications. J. Chem. Educ. **60**(1), A25 (1983)
3. Christensen, P.A., Hamnet, A.: Techniques and Mechanisms in Electrochemistry. Blackie Academic & Professional, Glasgow (2007)
4. Bard, A.J., Faulkner, L.R., White, H.S.: Electrochemical Methods: Fundamentals and Applications. Wiley, Hoboken (2022)
5. Oldham, K., Myland, J.: Fundamentals of Electrochemical Science. Academic Press, San Diego (2012)
6. Survila, A.: Electrochemistry of Metal Complexes: Applications from Electroplating to Oxide Layer Formation. Wiley, Hoboken (2015)
7. Callister, W.D., Rethwisch, D.G.: Materials Science and Engineering: An Introduction. Wiley, New York (2018)
8. Karan, S., Banerjee, D., Datta, A., Goswami, A., Majumder, D.D.: Shape based characterization of nanoparticles–a fuzzy mathematical approach. Proc. Indian Natn. Sci. Acad. **81**(5), 1183–1192 (2015)
9. Ren, X., Wang, H., Wei, Q., Lu, Y., Xiao, B., Xie, J.: Electrochemical behaviour of N80 steel in CO_2 environment at high temperature and pressure conditions. Corros. Sci. **189**, 109619 (2021)
10. Li, K., Zeng, Y., Luo, J.L.: Influence of H_2S on the general corrosion and sulfide stress cracking of pipelines steels for supercritical CO_2 transportation. Corros. Sci. **190**, 109639 (2021)
11. Wang, Q., Wu, W., Li, Q., Zhang, D., Yu, Y., Cao, B., Liu, Z.: Under-deposit corrosion of tubing served for injection and production wells of CO_2 flooding. Eng. Fail Anal. **127**, 105540 (2021)
12. Santos, V.H.J.M.D., Pontin, D., Ponzi, G.G.D., Stepanha, A.S.D.G.E., Martel, R.B., Schütz, M.K., Einloft, S.M.O., Vecchia, F.D.: Application of Fourier Transform infrared spectroscopy (FTIR) coupled with multivariate regression for calcium carbonate ($CaCO_3$) quantification in cement. Constr. Build. Mater. **313**, 125413 (2021)
13. Garg, N., Wang, K., Martin, S.W.: A Raman spectroscopic study of the evolution of sulfates and hydroxides in cement–fly ash pastes. Cem. Concr. Res. **53**, 91–103 (2013)
14. Choudhary, H.K., Anupama, A.V., Kumar, R., Panzi, M.E., Matteppanavar, S., Sherikar, B.N., Sahoo, B.: Observation of phase transformations in cement during hydration. Constr. Build. Mater. **101**, 122–129 (2015)
15. Bronnikov, A., Killian, D.: Cone-beam tomography system used for non-destructive evaluation of critical components in power generation. Nucl. Instrum. Methods. Phys. Res. B. **422**(1–3), 909–913 (1999)
16. Miao, X., Zhang, L., Wang, Y., Wang, L., Fu, X., Gan, M., Li, X.: Characterisation of wellbore cement microstructure alteration under geologic carbon storage using X-ray computed microtomography: a framework for fast CT image registration and carbonate shell morphology quantification. Cem. Concr. Compos. **108**, 103524 (2020)
17. Gan, M., Zhang, L., Miao, X., Oladyshkin, S., Cheng, X., Wang, Y., Shu, Y., Su, X., Li, X.: Application of computed tomography (CT) in geologic CO_2 utilization and storage research: a critical review. J. Nat. Gas. Sci. Eng. **83**, 103591 (2020)
18. Akin, S., Kovscek, A.R.: Computed tomography in petroleum engineering research. Geol. Soc. Spec. Publ. **215**(1), 23–38 (2003)
19. Fang, G., Dong, B., Han, N., Xing, F.: Investigation of a novel 3D non-destructive evaluation for corrosion process in reinforced concrete. In: 5th International Conference on Durability of Concrete Structure, 2016 ICDCS, Shenzhen, China, Corrosion 7, pp. 296–299. Purdue University Press (2016)
20. Xue, Q., Zhang, L., Mei, K., Wang, L., Wang, Y., Li, X., Cheng, X., Liu, H.: Evolution of structural and mechanical properties of concrete exposed to high concentration CO_2. Constr. Build. Mater. **343**, 128077 (2022)

Chapter 3
Corrosion in CO_2 Capture and Transportation

Shijian Lu and Liwei Zhang

3.1 Corrosion in CO_2 Capture

CO_2 capture is a process that captures CO_2 emissions from sources like coal-fired power plants [1]. CO_2 capture can be classified into three categories [2]: (1) CO_2 is captured as a pure or near-pure CO_2 stream, either from an existing industrial process or by modifying the existing process to generate such a stream (e.g., pre-combustion fuel gasification, oxyfueling power generation, etc.; (2) CO_2 comes from flue gas with relatively low CO_2 concentration, and CO_2 is concentrated into a pure or near-pure CO_2 stream (e.g., post combustion separation from power plant or cement plant flue gases); (3) CO_2 is directly captured from the air to form a pure CO_2 stream or into a chemically stable end product. Among the three categories, post-combustion CO_2 capture technology is regarded as the most mature technology—a number of post combustion capture systems have been in commercial operation since the late 1970s, providing CO_2 for refrigeration and food-processing purposes. These installations provide pilot-scale proof of post-combustion CO_2 capture technology [2].

In post-combustion CO_2 capture, amine-based absorbents are the most-widely used CO_2 capture agent, which has been broadly tested and validated in the field of natural gas purification. A typical amine-based CO_2 absorption process can be described as follows. The lean amine solvent (with a low content of CO_2 reaction

S. Lu (✉)
Carbon Neutrality Institute, China University of Mining and Technology, Xuzhou 221008, Jiangsu, China
e-mail: lushijian@cumt.edu.cn

Jiangsu Key Laboratory of Coal-Based Greenhouse Gas Control and Utilization, China University of Mining and Technology, Xuzhou 221008, Jiangsu, China

L. Zhang
State Key Laboratory of Geomechanics and Geotechnical Engineering, Institute of Rock and Soil Mechanics, Chinese Academy of Sciences, Wuhan 430071, China

University of Chinese Academy of Sciences, Beijing 100049, China

L. Zhang (ed.), *Corrosion in CO_2 Capture, Transportation, Geological Utilization and Storage*, Engineering Materials, https://doi.org/10.1007/978-981-99-2392-2_3

products) is brought into contact with cooled flue gas in a packed absorber tower. In the absorber tower, CO_2 in the flue gas is absorbed by the lean amine solvent, resulting in removal of CO_2 from the flue gas and enrichment of CO_2 in the amine solvent. The amine solvent with a high content of CO_2 reaction products (hereafter referred to as the rich amine solvent) exits the base of the absorber tower and is pumped to the top of the amine stripping tower. In the amine stripping tower, the rich amine solvent is heated so the CO_2 absorption reaction is reversed, releasing pure CO_2 and regenerating the lean amine solvent. The released pure CO_2 exits the top of the stripping tower and is collected for CO_2 transportation to the storage site. The lean solvent from the bottom of the stripping tower is cooled and cycled back to the absorber [2]. After years of research and development, the energy consumption to re-generate amine-based absorbents has been reduced to 2.35 GJ/t CO_2, which makes CO_2 capture commercially competitive under the promotion of incentives like Section 45Q tax credit of the United States Internal Revenue Code. Among different types of amines for post-combustion CO_2 capture, monoethanolamine (MEA) is relatively competitive due to its outstanding performance. Thanks to MEA's good absorption capacity for CO_2 and a fast reaction rate between MEA and CO_2, the height of the absorption column and the investment of the equipment can be reduced [3].

Corrosion is considered to be one of the most severe operational problems in amine-based CO_2 absorption processes and experience shows that amine degradation products often aggravate corrosion. In general, amines are not intrinsically corrosive, since they associate both high pH and low conductivity. However, amines may become corrosive when they absorb CO_2. Furthermore, since the treatment units operate in semi-closed circuit, the solvent may become enriched with possibly corrosive degradation products [4]. The main materials of CO_2 absorption towers and desorption towers in CO_2 capture are carbon steel such as A36 and A106, and stainless steel such as 302, 304, 316 and 410. Amines undergo thermal, oxidation, and chemical degradation during CO_2 absorption and desorption, creating the potential for chemical corrosion. Corrosion has always been an issue in natural gas sweetening using amines. Some forms of corrosion include general corrosion, erosion corrosion, stress corrosion cracking, pitting, crevice, and galvanic corrosion [5, 6]. Besides amine-induced corrosion, flue gas rich in CO_2 and water can cause corrosion of pipelines and flue gas pre-treatment units, due to the formation of carbonic acid as a result of CO_2 dissolution in water.

Amine degradation is an important contributor to corrosion in post-combustion CO_2 capture plants. O_2-induced oxidative degradation and thermal degradation at higher temperature with CO_2 in the stripper or at the hot end of a lean-rich heat exchanger are the two main types of degradation [7, 8]. During the amine degradation process, nonregenerable heat-stable salts (HSSs) with a considerable amount are produced, and the accumulation of these salts can cause severe corrosion of the equipment [3]. Temperature has a noticeable effect on corrosion phenomena in the CO_2 capture process since most electrochemical reactions involved are thermally activated. The corrosion rate is generally doubled when the operating temperature increases by 10–20 °C. In the CO_2 capture process, temperatures vary widely in different units, ranging from 40 °C in the absorber up to 130 °C in the reboiler [4]. Taking into account both the CO_2 loading and the temperature, the main corrosion risks are usually encountered in areas with high CO_2 loading and high temperatures [9]. These conditions are generally found in the rich amine line after the heat exchanger and up to the regenerator input. The presence of impurities like H_2S and NO_x also contributes to corrosion in the CO_2 capture process. H_2S both causes direct corrosion of steel and toxicates the amine solution. the presence of NO_x accelerates the overall degradation rate of amines [10] and may cause a high corrosion rate of the steel at the initial stage of CO_2 absorption by amine [3]. The type of amine is also an important factor governing the rate of corrosion. Usually, primary amines (e.g. MEA) are the most corrosive, secondary amines (e.g. DiEthanolAmine, DEA) are slightly less corrosive, and tertiary amines (e.g. MethylDiEthanolAmine, MDEA) exhibit the lowest risk of corrosion [11, 12]. Amine concentration also has an influence on corrosion. Excessively high amine concentrations should generally be avoided [4]. Finally, a high solvent flow rate and the occurrence of turbulent flow may cause risks of erosion-corrosion. Erosion-corrosion is common for carbon steels and is relatively uncommon for stainless steel, since stainless steel grades are far more resistant to erosion. When the content of degradation products becomes too high, the resistance of carbon steels to erosion-corrosion becomes weak, and the degree of erosion-corrosion for carbon steels is noticeably aggravated. Some of the degradation products have a chelating effect on iron and may cause fast dissolution of the protective deposits [13, 14].

To improve the knowledge of corrosion processes in amine units, studies have been carried out since 1990s to investigate the mechanism and the level of corrosion in amine solution [15]. Muhammad and Faical [16] investigated the corrosion phenomenon of carbon steel in amine and ionic liquid blends comprised of alkanolamines (monoethanolamine, 2-amino-2-methyl-1-propanol, diethanolamine, N-methyldiethanolamine) and hydrophilic ionic liquid ([BMIM][BF4], [EMIM][BF4], and [EMIM][Otf]) at room temperature. The effect of amine type, ionic liquid type, process temperature, CO_2 loading, presence of oxygen in flue gas, as well as the influence of water content on the corrosion rate of carbon steel was systematically studied [17]. Xiang et al. [18] studied the effects of different control factors (O_2, HSSs, flow and temperature) on the corrosion of A36 carbon steel in 30 wt% MEA solution, and the results revealed that HSSs resulted in a higher corrosion rate at the initial stage, but had a relatively small effect on the corrosion rate for longer exposure time. Campbell et al. [19] evaluated the corrosive behaviour of loaded amine solvents under stripper operating conditions, for post-combustion carbon capture, to determine the feasibility of using carbon steel in CO_2 capture plant construction. Four amine solvents (i.e., monoethanolamine (MEA), methyldiethanolamine (MDEA), 2-amino-2-methyl-1-propanol (AMP), and 1-(2-aminoethyl)piperazine (AEPZ)) were studied when in contact with carbon steel (C1018) over a 28-day reaction period. They found that MDEA and AMP can preferentially produce sufficient surface $FeCO_3$ layers to reduce corrosion levels in carbon steels for use under stripper conditions in post-combustion carbon capture plants. However, MEA and AEPZ show significant corrosion to the carbon steel surface. Xiang et al. [3] comparatively investigated the effect of NO_3^-

on the corrosion behaviour of A106 carbon steel in fresh and dirty MEA solutions during the CO_2 capture process. They found that adding 11,000 ppmw $NaNO_3$ has a distinct effect on the corrosion of A106 carbon steel at the initial stage, and has a minimal effect on the corrosion at the later stage.

Previous studies have revealed that the corrosion rate depends on the materials used in the CO_2 capture plant. Stainless steel is more resistant to corrosion than carbon steel and most of the key components of the CO_2 capture plants are now constructed with the use of stainless steel. Degradation of amines can be oxidative or thermal and some of the degradation products are corrosive agents and inevitably cause equipment corrosion [6], but the stainless steel has a good resistance to corrosion induced by amine degradation products [4]. Studies have recommended to use stainless steel in both the amine-rich sections of the unit and the lean solvent sections of the unit [4, 20, 21], in order to operate at higher flow rates, and also to provide versatility to the unit for easier solvent swapping.

Another alternative for corrosion protection is corrosion inhibitors. Use of corrosion inhibitors is often recommended when the operator would like to minimize investment costs and to use carbon steel as the building material for most components. Two types of corrosion inhibitors are commonly used. Oxidizing passivators react electrochemically with the steel surface to promote the formation of a stable and protective passive layer. In most cases, the oxidizing passivators consist of inorganic compounds. Representative oxidizing passivators include sodium metavanadate, salts of heavy metals like antimony, cobalt, bismuth or nickel, and copper salts [4]. Film forming inhibitors represent the other family of molecules commonly used for corrosion inhibition applications. This type of molecules adsorbs on the steel surface to form a thin layer, which impedes access of the corrosive solution to the surface. However, for CO_2 capture applications, they are considered to be less efficient than inorganic oxidizing passivators [22].

3.2 Corrosion in CO_2 Transportation

3.2.1 General Introduction

CO_2 transportation is an important stage in whole-chain CCUS projects. CO_2 transportation technology is to transport the CO_2 captured at the emission sources to the place of CO_2 utilization or CO_2 storage through specific transporting approaches (such as tanker trucks, ships, pipelines, etc.). A scientific design of the CO_2 transportation plan is the key to match CO_2 sources with the corresponding sinks, and is the key to maintain a stable operation of the whole-chain CCUS system [23]. In order to improve efficiency, CO_2 is generally transported under high pressure, so it is necessary to increase the pressure of CO_2 during transportation. The main means of CO_2 transportation include pipeline transportation, truck transportation, train transportation and ship transportation. Compared with truck transportation, train transportation and ship transportation, pipeline transportation has the lowest per unit cost and is ideal for large-scale CO_2 transportation. Specifically, the costs of truck transportation, train transportation and ship transportation are all higher than 0.14 USD/(tCO_2·km), while pipeline transportation cost for supercritical CO_2 is 0.06 ~ 0.07 USD/(tCO_2·km). In the future, with the increase of transportation scale, it is expected that the cost of pipeline transportation can be reduced to less than 0.04 USD/(tCO_2·km).

In the 1970s, in order to improve the recovery rate of crude oil, the United States began to use pipelines to transport pure CO_2 to oilfields and use CO_2 for the purpose of CO_2-enhanced oil recovery (CO_2-EOR). In 1972, Canyon Reef Carriers (CRC) completed and commissioned the first CO_2 pipeline to transport natural CO_2 to the Texas SACROC oilfield. With more than 50 years of technology development, pipelines have been widely used in developed countries for long-distance and large-scale transportation of CO_2. Nowadays, the construction of CO_2 transportation pipeline networks that match the needs of large-scale whole-chain CCUS projects is going on, and the total length and total transportation capacity of CO_2 pipelines have rapidly increased. There are about 7000 km of CO_2 pipelines in the world with a total capacity of 150 Mt/a, most of which are located in the United States, Canada, Norway and Turkey, with at least 22 CO_2 commercial pipeline projects in operation [23]. Due to the cost advantages of supercritical phase CO_2 transportation, most of the commercial CO_2 pipeline transportation projects transport CO_2 in supercritical phase. During CO_2 transportation in pipelines, compressors at the upstream end provide the driving force to move CO_2, and some pipelines are equipped with intermediate pressure booster stations to maintain the transport efficiency of CO_2.

Nearly 90% of the existing CO_2 transportation pipelines are for CO_2-EOR, and some examples are discussed as follows. Sheep Mountain CO_2 transportation pipeline was built in 1983, with a total pipeline length of 656 km. The original 296 km pipeline had a diameter of 20 inch with a designed transport capacity of 6.3 Mt/year, and the later-built 360 km pipeline had a diameter of 24 inch with a designed transport capacity of 9.2 Mt/year. Sheep Mountain Pipeline System is transporting CO_2 from Colorado and New Mexico to West Texas, USA, helping recover up to 25% of the remaining oil in Permian Basin oil fields. The Cortez CO_2 pipeline stretches approximately 808 km from the McElmo Dome Field in Colorado to Denver City, Texas, with the purpose to transport large amounts of CO_2 to Wasson Oilfield for CO_2-EOR. The pipeline transports around 19.3 Mt of CO_2 per year, with a diameter of 30 inch. The pipeline passes through 2 metering stations, 3 pressure reducing stations and 1 pumping station to reach the Denver City metering station. The designed CO_2 transport pressure is 9.6–18.6 MPa, and the maximum transport temperature is 43 °C. Driven by President Obama's stimulus package, several CO_2-EOR projects were launched in the United States in 2010s, and a $500 million-cost pipeline to annually transport millions of tons of CO_2 from Great Plains Synfuels Plant in Antelope Valley, North Dakota, USA to Weyburn oilfield, Saskatchewan, Canada, was completed in 2012 [11]. This pipeline network is part of Weyburn CO_2 Enhanced Oil Recovery Project, which utilizes North Dakota Gasification Company's CO_2 from methane production to extract oil by injecting CO_2 into oil reservoirs in Weyburn oilfield.

In China, the CO_2 transport project with the longest pipeline is Sinopec's Qilu-Shengli CCUS Demonstration Project. This project intends to transport CO_2 from Qilu Petrochemical Plant to Shengli Oilfield in Zibo City, Shandong Province, China, with the target to reach 1 Mt/year CO_2 transport capacity. Information about some long-distance CO_2 transportation pipelines worldwide is shown in Table 3.1 [23].

Table 3.1 Information on some long-distance CO_2 transportation pipelines worldwide

Pipeline name	Country	Operator	CO_2 transport capacity (Mt/a)	Length (km)	Pipeline diameter (inch)	CO_2 source
Cortez	USA	Kinder Morgan	19.3	808	30	Natural CO_2 in subsurface
Sheep Mountain	USA	Occidental, BP	9.5	660	20 and 24	Natural CO_2 in subsurface
Bravo	USA	Occidental, BP	7.3	350	20	Natural CO_2 in subsurface
Canyon Reef Carriers	USA	Kinder Morgan	5.2	225	16	Gasification plant
Val Verde	USA	Occidental	2.5	130	–	Gas treating plant
Bati Raman	Turkey	Turkish Petroleum Corporation	1.1	90	–	Dodan Oilfield
Weyburn	USA, Canada	Dakota Gasification Company	5	328	12 ~ 14	Gas treating plant
NEJD	USA	Denbury Resources	–	295	20	–
Transpetco Bravo	USA	Transpetco	3.3	193	12.75	–
Snøhvit	Norway	StatoilHydro	0.7	153	8	–
West Texas	USA	Trinity	1.9	204	8 ~ 12	–
Este	USA	Exxon Mobil	4.8	191	12、14	–
Central Basin	USA	Kinder Morgan	11.5	–	16 and 26	–
SACROC	USA	–	4.2	354	16	–
Qilu-Shengli	China	Sinopec	1	75	–	Petrochemical plant
In Salah	Algeria	–	–	14	–	–
Reconcavo	Brazil	–	–	183	–	–
Lacq	France	–	–	27	8–12	–
Barendrecht	Netherland	–	–	20	14	–

Note "–" represents "unknown information"

In a pipeline, CO_2 can be transported in gas phase, liquid phase, and supercritical phase. Supercritical CO_2 transport is preferred for large-scale industrial long-distance CO_2 pipelines. No matter which CO_2 phase exists in the pipeline, a single-phase CO_2 transport needs to be ensured during the transportation process. A co-existing phase transportation like a gas phase mixing with a liquid phase is not desired, because the two-phase transportation may cause blocking of one phase by the other phase in the pipeline transportation process [22], which has a serious impact on the transportation capacity of the pipeline.

Low-pressure gas-phase CO_2 transportation is generally used in short distance and small flow rate CO_2 transport. The pressure for gas-phase CO_2 transportation usually does not exceed 4 MPa, and an insulation layer may need to be installed surrounding the pipeline. The gas-phase CO_2 transportation is uneconomical and is not suitable for large flow rate CO_2 transport. Liquid-phase or supercritical phase CO_2 transportation is ideal for large flow rate CO_2 transport. In liquid-phase CO_2 transportation, CO_2 stays in liquid state in the pipeline, and the transportation pressure is increased by pumping to overcome the friction and gravity forces along the way. Depending on the temperature condition, an insulation layer may need to be installed surrounding the pipeline. Liquid-phase CO_2 transportation delivers much more CO_2 than gas-phase CO_2 transportation given the same pipe diameter. In supercritical-phase CO_2 transportation, the temperature and pressure in the pipeline are higher than those of the critical values for supercritical CO_2, so CO_2 is maintained in a supercritical state in the pipeline. The transportation pressure is raised and maintained by a compressor. In most cases, the pipeline does not need to be insulated. During CO_2 transportation, the CO_2 phase density varies with the change of ground temperature, so the intermediate booster station needs to choose the appropriate pressurization level according to the CO_2 density. Supercritical CO_2 transport can maintain a high transport capacity, and also has a much less flowing resistance than the gas phase and the liquid phase, which helps maintain pumping efficiency. Therefore, supercritical phase is the preferred CO_2 state in the transport pipeline.

3.2.2 Impurities in CO_2 Transportation and Their Role in Pipeline Corrosion

Impurity is an important factor in pipeline corrosion by CO_2, because pure CO_2 is not corrosive and does not cause pipeline corrosion. H_2O, N_2, NO_x, O_2, H_2, CH_4, H_2S, Ar, etc. are typical impurities in pipeline CO_2 transportation. The properties of CO_2 containing impurities such as N_2 and O_2 determine the parameters of phase equilibrium, density, compressibility, viscosity, etc. [22], which influence a series of system parameters like the pressure curve/loss conditions, heat transfer, flow volume, density curve, operating capacity limit, hydrate formation, etc. in the pipeline transportation process. The existing models can well describe the properties, equations of state (EOS) and transport characteristics of pure CO_2. However, those models need to be improved to better describe the properties, EOS and transport characteristics of CO_2 containing impurities [24–27]. Sometimes it is necessary to have experimental results to correct the model-predicted results. For CO_2 transport with impurities, the transport properties such as viscosity and density need to be derived from EOS considering impurities. The content and composition of impurities have a great influence on the phase diagram of CO_2 [26, 28], which may increase the critical pressure of CO_2 and decrease the critical temperature, and reduce the supercritical phase area on the phase diagram, so that the optimal pipeline operation pressure and flow rate need to be adjusted. At present, for CO_2 with impurities, it is necessary to conduct in-depth studies on the influence of intermolecular forces and molecular shape on the phase characteristics of CO_2-impurity mixtures, so as to improve the model accuracy of the physical property parameters and state equations of impure CO_2.

The presence of impurities in CO_2 has two adverse effects on the transportation and safe operation of pipelines [23]. First, the presence of impurities changes the boundaries of the two-phase zone, which may cause two-phase co-flow and free water formation and subsequently lead to hydrate formation and corrosion problems. The presence of impurities also requires a higher pipeline operating pressure. Second, impurities may occupy available spaces in the pipeline and affect the CO_2 compressibility, resulting in a decrease in pipeline transportation capacity. Additional energy is needed for CO_2 compression during CO_2 transportation, and as the amount of impurities increases, the transportation capacity of the pipeline decreases. Studies have shown that a presence of 5% and 10% CH_4 in CO_2 flow reduces the CO_2 transport capacity by 9.4% and 16%, respectively, compared with the transport of pure CO_2 under the same conditions. If 5% N_2 is present, the CO_2 transport capacity of the pipeline will be reduced by 12.6%. Therefore, the total content of impurities needs to be restricted, which avoids occupation of pipeline volume and an increase in investment and energy consumption. A recommendation of concentration limits for impurities is provided in Table 3.2.

H_2O is an important impurity that causes corrosion in CO_2 pipelines. Before the CO_2 enters the pipeline system, the CO_2 must be dewatered to reduce H_2O content. The impact of H_2O on the CO_2 transport pipeline can be described in the following two scenarios. Scenario I is a corrosive environment with CO_2 as the main body containing a small amount of H_2O and other impurities, and the corrosion level of this environment on carbon steel pipes is related to the control of water and impurity contents. Typical corrosive impurities include H_2S, O_2, H_2, etc. A combination of H_2S or O_2 with H_2O can cause noticeable corrosion, even the H_2O content is tiny. the concentration of H_2S in the fluid stream must be kept below the solubility limit to prevent H_2S corrosion. On the other hand, a presence of H_2 from precombustion capture may cause hydrogen embrittlement [2]. Scenario II is that the water content in the pipeline is relatively large, and the corrosion of free water occurs. When the water content is greater than the critical water content, there is a risk of fast corrosion forming pinholes in the pipeline. The corrosion of free water is a corrosion induced by CO_2-saturated water (forming carbonic acid) on carbon steel pipes, and the corrosion rate of carbon steel pipes in this scenario is very high. As a result, corrosion-resistant

Table 3.2 Concentration limits of impurities for CO_2 transportation and storage

Component	Concentration limit	Consideration
H_2O	500 ppm	Prevention of free water
H_2S	200 ppm	Minimization of pipeline corrosion; health and safety
CO	2000 ppm	Health and safety
SO_x	100 ppm	Minimization of pipeline corrosion; health and safety
NO_x	100 ppm	Minimization of pipeline corrosion; health and safety
O_2	< 4 vol%	Minimization of pipeline corrosion; optimal for aquifer storage
	< 1000 ppm	Technical limit, for EOR
CH_4	< 4 vol%	Optimal for aquifer storage
	< 2 vol%	For EOR
$N_2 + Ar + H_2$	< 4 vol% total	Lower for H_2 in view of hydrogen embrittlement and economic value of its energy content

materials are required. In general, corrosion resistant alloys with a Cr content higher than 13% have good corrosion resistance in the corrosive environment containing free water [29]. An inclusion of a corrosion allowance in the design wall thickness of the pipeline is also feasible [2].

In Scenario I, the corrosion inhibitors commonly used in oil and gas fields such as imidazoline, cetyl succinate acyl, cetyl trimethyl ammonium bromide, etc. can reduce the corrosion rate and inhibit formation of pits in carbon steel exposed to CO_2 and H_2O. However, the corrosion cannot be completely stopped for carbon steel, and the corrosion rate of carbon steel is usually higher than 1 mm/year [30]. The development of new corrosion inhibitors that are compatible with both water and CO_2 and can induce formation of a protective layer on carbon steel surface will be a research direction for pipeline corrosion control in CO_2 transport conditions. In Scenario II, the presence of free water causes a high corrosion rate, and corrosion inhibitors only are not sufficient to reduce the corrosion rate. Therefore, it is necessary to use stainless steel containing Ni and high Cr content as the material of CO_2 transportation pipelines, with consideration of both cost and corrosion resistance. The development of stainless steel coatings and optimal inclusion of anti-corrosion additives in stainless steel are hot research directions of CO_2 transportation pipeline corrosion control. In the presence of free water, CO_2 can also cause damage to pipe fittings and sealing materials such as pipe seats, gaskets, and O-rings [31]. The rubber commonly used in the oil and gas industry cannot meet the requirements of CO_2 transportation pipeline. Special rubbers for high-pressure CO_2 pipeline systems such as PTFE, polychlorotrifluoroethylene, polyetheretherketone resin, acid amine fiber (nylon), ethylene-propylene polyphthalimide, semi-rigid polyurethane and kalrez can be considered [32].

At present, the impact of CO_2 flowing process, water content, and the interaction of water and other impurities on pipeline corrosion is a hot research topic. Previous studies have reported that a water content of less than 80% of water saturation in CO_2 does not cause obvious local corrosion of carbon steel pipeline and only uniform corrosion is observed, and the uniform corrosion rate is less than 0.1 mm/year [33, 34]. In fact, the water content in CO_2 is generally much lower than 80% of water saturation in CO_2 in most CO_2 transport pipelines. For example, most CO_2 transport pipelines in the U.S. require that the water content is below 50 ppmv. Compared with other projects, the Weyburn project has a stricter water content restriction. The water content in the CO_2 gas stream is strictly limited to 20 ppmv in the Weyburn pipeline. Though the CO_2 gas stream contains 9000 ppmv H_2S, no significant corrosion has occurred since the start of the project, due to successful control of water content in the CO_2 gas stream [35]. In China, PetroChina Jilin Oilfield uses carbon steel pipe lined with 2 mm 316L stainless steel to transport wet CO_2 with 50–100 ppmv water content, and no pipeline damage caused by corrosion has been reported. Another issue that needs to be considered is the formation of CO_2 hydrates when water is present. The formation of CO_2 hydrates can cause blockage in pipes and even damage the pipeline and the equipment. In order to reduce corrosion, the water content should be maintained below 200 ppm to avoid formation of CO_2 hydrates.

If defects occur on the pipeline due to corrosion, CO_2 leakage accidents from the pipeline may occur during long-distance pipeline transportation. Other causes that can cause defects on the pipes include mechanical damage, soil movement, material quality, etc. Because CO_2 gas has a higher density than the air, CO_2 is easy to accumulate in the low altitude area, which will cause harm to people, animals and plants in the low altitude area. Moreover, the pressure, temperature and the phase of CO_2 in the leakage process are changing, and the temperature drop often causes the fracture toughness of the pipeline to decline sharply. As a result, brittle fracturing of the pipeline is easy to occur and fracture propagation can become a serious problem. Fracture propagation is very likely to cause hundreds of kilometers of pipeline damage in a short time when a pipeline leaking occurs. The thinning of CO_2 pipes in service due to the erosion of the external environment and the internal CO_2 corrosion will significantly increase the risk of leakage accidents. In addition, during the production and installation of the pipeline, it is inevitable to have dents, cracks and other defects in the pipeline. These defects serve as leakage starting sources and the sources can be expanded by corrosion damage of the pipeline, eventually leading to heavy leakage of the pipeline and fracturing damage. The leakage and fracturing damage are often without warning, and once they occur, the consequences are serious.

3.2.3 Guides for CO_2 Transportation System Design

Safe management of CO_2 pipeline transportation (including corrosion control) and development of CO_2 leakage monitoring technology are key guarantees to achieve large-scale industrial application of CO_2 transportation. However, there are no world-unified CO_2 pipeline transportation standards. The U.S. has extensive engineering experience in CO_2 pipeline transportation [24], but the CO_2 pipeline design and operation standards in the U.S. simply follow oil and gas pipeline transportation standards, and no comprehensive standards for CO_2 pipeline transportation are made. Canada and Australia have similar situations to the U.S. Norway is the only country that has a specialized CO_2 pipeline design standard. DET NORSKE VERITAS (DNV) in Norway developed the CO_2 Pipeline Design Code (DNV-RP-J202) in 2010.

In summary, most countries do not have specialized CO_2 pipeline standards. The relevant design standards in oil and gas industry like ASME B31.8, ASME 831.4, IP6, BS EN 14161, BS PD 8010, DNV-OS-F101, etc. play a very important role in the construction of CO_2 transport pipelines. However, those standards and codes need to be further supplemented in the aspects of process conditions, pipeline design, safety and environment, material selection, cleaning and strength testing, construction technology, pipeline operation, leakage monitoring, etc. Also, Those standards may not be applicable for pipeline design in densely populated areas [23]. The design of the pipeline needs to consider a series of designing parameters like the pipe diameter, pipe length, specific gravity, density, compression coefficient, viscosity, heat capacity containing impure CO_2, operation temperature, operation pressure, friction coefficient, etc. A combination of the main CO_2 transport line and branch lines increases the difficulty of design. A successful design needs to ensure a steady and single-phase flow. The deployment of the pressurization station is related to the pressure and CO_2 phase state in the pipeline. Due to the height difference and friction loss, the pressure in the pipeline may rise or drop, which leads to the phase state change of CO_2. As a result, it is necessary to build intermediate pressurization stations to adjust the pressure to maintain single-phase CO_2 flow in the pipeline. An optimization of the distance between pressurizing stations effectively reduces energy consumption [32, 36].

Strict requirements should be enforced for the control of minor defects in pipeline materials at the time of design. Under the combined effects of high pressure, temperature change and chemical corrosion, ductility cracking may occur in the pipeline, which is one of the serious safety risks during pipeline operation [37]. As a result, there must be corresponding material fracture toughness requirements to achieve fracture control. If the cracking behavior of the pipeline steel cannot be controlled by applying appropriate pipeline material, installation of crack relief devices can be considered. The crack relief device is used to prevent pipeline cracking. The crack relief device is typically composed of glass fiber wrapped around the pipe and embedded in an epoxy resin, and sometimes steel hoops are also used [2]. For long-distance CO_2 pipelines, crack relief devices can be set up at pipeline intervals to ensure that even if a break occurs, the affected length of the pipeline is minimized, and the loss and maintenance costs are reasonable [38, 39]. In the CO_2 pipeline systems operating in the U.S., it is common practice to install crack relief device to impede the propagation of axial fractures [2].

For sub-sea CO_2 pipeline transportation, no specified standards or regulations have been made. Two standards–Code of practice for pipelines-part 4: risk-based integrity management of steel pipelines on land and subsea pipelines (BS PD 8010-4) and Submarine pipeline systems (offshore standard DNV-OS-F101) cover some information about sub-sea pipeline system design, but the two standards are not specially made for CO_2 transport pipeline. In terms of technical and economic analysis, based on the technical characteristics and cost analysis methods of existing CO_2 submarine pipelines, technical and economic models of submarine CO_2 pipelines have been established. Those models serve as guides for technical design and cost analysis of the CO_2 transportation part of offshore CO_2 storage projects. Some studies have concluded that the factors to be considered in the design of on-shore and subsea pipelines are quite similar, so subsea pipelines can be designed, constructed and managed in accordance with the relevant standards and specifications for onshore pipelines [40]. However, the cost of constructing subsea pipelines is significantly higher than that of the on-shore pipelines, regular maintenance of subsea pipelines is more difficult than that of the on-shore pipelines, and the handling of leakage accidents for subsea pipelines is more challenging than that of the onshore pipelines. Therefore, it is necessary to develop specific standards to regulate the design and construction of subsea CO_2 transport pipelines.

In summary, there is no systematic standard and specification for the design and supervision and inspection of CO_2 pipeline transmission within the framework of international standards, and there is no guiding principle for CO_2 pipeline transmission. Although the 22 sets of large-scale commercial CO_2 pipeline designs documented internationally are based on the corresponding national standards [23], these standards (ASME B31.8, ASME 831.4, IP6, BS EN 14161, BS PD 8010, DNV-OS-F101, etc.) are general standards for pipeline transportation, and only DNVRP-J202 (CO_2 Pipeline Design and Operation Manual) is explicitly for CO_2 pipeline transportation. Most of the existing standards and norms need to be further improved and cannot be fully applied to CO_2 pipeline transportation. The literature clarifies that existing standards need to be improved and supplemented in the following aspects before they can be used for CO_2 pipeline transportation: process conditions, pipeline design, safety and environment, material selection, cleaning and strength testing, construction technology, pipeline operation, measurement, etc. In general, in order to ensure the safety and reliability of CO_2 pipelines, whether it is a national standard, industry standard or international standard, it is necessary to give the design principles and calculation methods of safety distance, spacing of globe valves, anti-corrosion, CO_2 leakage blocking, and safety distance. In addition, standards for subsea CO_2 pipelines are still missing. Therefore, there is an urgent need to formulate comprehensive and relatively unified risk and uncertainty standards and guidelines for pipeline CO_2.

In summary, there is no specific standard for the design, supervision and inspection of CO_2 transport pipeline within the framework of international standards. Although the 22 large-scale commercial CO_2 pipeline transportation projects documented are based on the corresponding national standards of the project locations [23], these standards (i.e., ASME B31.8, ASME 831.4, IP6, BS EN 14161, BS PD 8010, DNV-OS-F101, etc.) are standards for pipeline transportation in general, and only DNVRP-J202 (CO_2 Pipeline Design and Operation Manual) is explicitly for CO_2 pipeline transportation. Most existing pipeline transportation standards need to be further improved for better application to CO_2 pipeline transportation. Existing standards need to be improved and supplemented in the following aspects before they can be used for CO_2 pipeline transportation: process conditions, pipeline design, safety and environment, corrosion control, material selection, cleaning, strength testing, construction technology, pipeline operation, measurement, etc. In general, in order to ensure the safety and reliability of CO_2 pipelines, it is necessary to include the design principles and calculation methods of safety distance, spacing of globe valves, anti-corrosion, CO_2 leakage blocking, safety distance, etc. in the corresponding standards. In addition, comprehensive standards for subsea CO_2 pipelines are still missing and need to be made. Therefore, there is an urgent need to formulate comprehensive and global-unified risk and uncertainty standards and guidelines for CO_2 pipeline transportation.

3.2.4 Summary

Pipeline transportation of CO_2 has been applied on a large scale and long distance in several countries, especially in the U.S.. The technical maturity of pipeline CO_2 transportation in the U.S. and Canada is high, and some European countries have made great breakthroughs in the construction and operation of CO_2 pipelines, and at least 22 CO_2 long-distance pipeline transport projects are running [23]. The existing problem is that there is no clear understanding of the safety and environmental impacts associated with pipeline CO_2 transportation containing impurities from different sources. The safety and environmental impacts of CO_2 pipeline transportation mainly refer to the impacts of pipeline corrosion, pipeline fracturing and pipeline seal failure, and the impacts of CO_2 leakage on the surrounding area. Another problem is that the existing experiences of CO_2 pipeline transportation are mostly for pipelines in sparsely populated areas, and those experiences are not suitable for new pipeline construction in densely populated areas. As a result, guidelines on the risks and uncertainties of pipeline CO_2 transportation are urgently needed.

In a whole-chain CCUS system, CO_2 pipeline transportation between CO_2 sources and sinks is a key section for achieving large-scale CO_2 emission reduction. The International Energy Agency (IEA) predicts that by 2050, the global CO_2 transport pipeline will reach 200,000 km. However, the CO_2 pipeline transportation technology reserve is still insufficient to meet the expected large-scale pipeline deployment, and further research in both fundamental science and applied engineering is required. Also, adequate industrial applications are needed to improve the maturity of the technology. Future technological developments should focus on the following aspects:

(1) Investigation on the material properties of the pipeline when supercritical phase CO_2 is transported;
(2) Design and optimization of multi-sink CO_2 pipeline system based on geographic data of the sinks;
(3) Development of an economic analysis method for CO_2 pipeline;

(4) Optimization of the pipeline transportation parameters for CO_2 transport containing impurities;
(5) Study on the influence of CO_2 transport process, water content, and other impurities on corrosion, research on corrosion control technology, and development of corrosion-resistant low-cost pipeline materials;
(6) Framework for pipeline transportation safety risk assessment;
(7) Make use of the existing experience in the design and construction of oil and gas pipelines to build a regulation system for CO_2 pipeline transportation.

References

1. U.S. Department of Energy: Carbon Capture, Utilization & Storage. https://www.energy.gov/carbon-capture-utilization-storage
2. Rackley, S.A.: Carbon Capture and Storage. Butterworth-Heinemann, Oxford (2017)
3. Xiang, Y., Xie, W., Ni, S., He, X.: Comparative study of A106 steel corrosion in fresh and dirty MEA solutions during the CO_2 capture process: Effect of $NO_3{}^-$. Corros. Sci. **167**, 108521 (2020)
4. Kittel, J., Gonzalez, S.: Corrosion in CO_2 post-combustion capture with alkanolamines–a review. Oil Gas Sci. Technol. **69**(5), 915–929 (2014)
5. Kohl, A.L., Nielsen, R.: Gas purification. Gulf Publishing Company, Houston (1997)
6. Fytianos, G., Ucar, S., Grimstvedt, A., Svendsen, H.F., Knuutila, H.: Corrosion evaluation of MEA solutions by SEM-EDS. ICP-MS and XRD, Energy Procedia **86**, 197–204 (2016)
7. Budzianowski, W.M.: Single solvents, solvent blends, and advanced solvent systems in CO_2 capture by absorption: a review. Int. J. Glob. Warm. **7**(2), 184–225 (2015)
8. Gouedard, C., Picq, D., Launay, F., Carrette, P. L.: Amine degradation in CO_2 capture. I. A review, Int. J. Greenh. Gas. Con. **10**, 244–270 (2012)
9. Bonis, M., Ballaguet, J., Rigaill, C.: A critical look at amines: a practical review of corrosion experience over four decades. In: 83rd Annual Convention of the Gas Processors Association 2004, San Antonio, U.S., vol. II, 644–659. GPA Midstream Association (2004)
10. Tanthapanichakoon, W., Veawab, A.: Heat stable salts and corrosivity in amine treating units. In: Proceedings of the 6th International Conference on Greenhouse Gas Control Technologies, Kyoto, Japan, vol. II, 1591–1594. Elsevier (2003)
11. DuPart, M., Bacon, T., Edwards, D.: Understanding corrosion in alkanolamine gas treating plants: part 1. Hydrocarb. Process. **72**(5), 6295319 (1993)
12. Veldman, R.: Alkanolamine solution corrosion mechanisms and inhibition from heat stable salts and CO_2. In: Corrosion 2000: National Association of Corrosion Engineers International Annual Conference and Corrosion Show, Orlando, U.S., NACE-00496. OnePetro (2000)
13. Nielsen, R., Lewis, K., McCullough, J., Hansen, D.: Corrosion in refinery amine systems. In: Corrosion 95: National Association of Corrosion Engineers International Annual Conference and Corrosion Show, Orlando, U.S., 128737. OnePetro (1995)
14. Al-Luqman, S., Al-Zahrani, A.M.: Methodology of mitigating corrosion mechanisms in amine gas treating units. In: Corrosion 2006: National Association of Corrosion Engineers International Annual Conference and Corrosion Show, San Diego, U.S., NACE-06441. OnePetro (2006)
15. Teevens, P.J.: Toward a better understanding of of the cracking behavior of carbon steel in alkanolamine sour gas sweetening units: its detection, monitoring and how to avoid it. In: Corrosion 90: National Association of Corrosion Engineers International Annual Conference and Corrosion Show, Houston, U.S., 198. OnePetro (1990)

16. Muhammad, H.R., Faical, L.: CO_2 capture in alkanolamine-RTIL blends via carbamate crystallization: route to efficient regeneration. Environ. Sci. Technol. **46**(20), 11443–11450 (2012)
17. Muhammad, H.R., Hana, B., Pascal, F., Mohamed, S., Faical, L.: Corrosion behavior of carbon steel in alkanolamine/room-temperature ionic liquid based CO_2 capture systems. Ind. Eng. Chem. Res. **51**(26), 8711–8718 (2012)
18. Xiang, Y., Yan, M.C., Choi, Y.S., Young, D., Nesic, S.: Time-dependent electrochemical behavior of carbon steel in MEA-based CO_2 capture process. Int. J. Greenh. Gas. Con. **30**, 125–132 (2014)
19. Campbell, K.L.S., Zhao, Y.C., Hall, J.J., Williams, D.R.: The effect of CO_2-loaded amine solvents on the corrosion of a carbon steel stripper. Int. J. Greenh. Gas. Con. **47**, 376–385 (2016)
20. Rooney, P.C., DuPart, M.: Corrosion in alkanolamine plants: causes and minimization. In: Corrosion 2000: National Association of Corrosion Engineers International Annual Conference and Corrosion Show, Orlando, U.S., NACE-00494. OnePetro (2000)
21. Kittel, J., Bonis, M., Perdu, G.: Corrosion control on amine plants: new compact unit design for high acid gas loadings. In: Sour Oil & Gas Advanced Technology Conference, vol. 27 (2008)
22. Bui, M., Adjiman, C.S., Bardow, A., Anthony, E.J., Boston, A., Brown, S., Fennell, P.S., Fuss, S., Galindo, A., Hackett, L.A.: Carbon capture and storage (CCS): the way forward. Energ. Environ. Sci. **11**(5), 1062–1176 (2018)
23. Onyebuchi, V.E., Kolios, A., Hanak, D.P., Biliyok, C., Manovic, V.J.R., Reviews, S.E.: A systematic review of key challenges of CO_2 transport via pipelines. Renew. Sust. Energ. Rev. **81**, 2563–2583 (2018)
24. Oosterkamp, A., Ramsen, J.: State-of-the-art overview of CO_2 pipeline transport with relevance to offshore pipelines. Haugesund: Research Council of Norway, Gassco and Shell Technology Norway (2008)
25. Chapoy, A., Nazeri, M., Kapateh, M., Burgass, R., Coquelet, C., Tohidi, B.: Effect of impurities on thermophysical properties and phase behaviour of a CO_2-rich system in CCS. Int. J. Greenh. Gas. Con. **19**, 92–100 (2013)
26. Munkejord, S., Hammer, M., Løvseth, S.: Intergovernmental panel on climate change, carbon capture & storage. Appl. Energy **169**, 499–523 (2016)
27. Porter, R.T.J., Fairweather, M., Pourkashanian, M., Woolley, R.M.: The range and level of impurities in CO_2 streams from different carbon capture sources. Int. J. Greenh. Gas. Con. **36**, 161–174 (2015)
28. Liljemark, S., Arvidsson, K., McCann, M.T., Tummescheit, H., Velut, S.: Dynamic simulation of a carbon dioxide transfer pipeline for analysis of normal operation and failure modes. Energy Procedia **4**, 3040–3047 (2011)
29. Zhang, Y., Gao, K., Schmitt, G.: Water effect on steel under supercritical CO_2 condition. In: Corrosion 2011: National Association of Corrosion Engineers International Annual Conference and Corrosion Show, Houston, U.S., NACE-11378. OnePetro (2011)
30. Zhang, Y., Gao, K., Schmitt, G.: Inhibiting steel corrosion in aqueous supercritical CO_2 conditions. Mater. Performance **50**(9), 54–59 (2011)
31. Schremp, F.W., Roberson, G.R.: Effect of supercritical carbon dioxide (CO_2) on construction materials. SPE J. **15**(03), 227–233 (1975)
32. Zhang, Z., Wang, G., Massarotto, P., Rudolph, V.: Optimization of pipeline transport for CO_2 sequestration. Energ. Convers. Manage. **47**(6), 702–715 (2006)
33. Jiang, X., Qu, D.R., Song, X.L., Liu, X.H., Zhang, Y.L.: Critical water content for corrosion of X65 mild steel in gaseous, liquid and supercritical CO_2 stream. Int. J. Greenh. Gas. Con. **85**, 11–22 (2019)
34. Xiang, Y., Wang, Z., Yang, X.X., Li, Z., Ni, W.D.: The upper limit of moisture content for supercritical CO_2 pipeline transport. J. Supercrit. Fluids. **67**, 14–21 (2012)
35. Xu, M., Zhang, Q., Yang, X., Wang, Z., Liu, J., Li, Z.: Impact of surface roughness and humidity on X70 steel corrosion in supercritical CO_2 mixture with SO_2, H_2O, and O_2. J. Supercrit. Fluids. **107**, 286–297 (2016)

36. Witkowski, A., Rusin, A., Majkut, M., Rulik, S., Stolecka, K.: Comprehensive analysis of pipeline transportation systems for CO_2 sequestration. Thermodynamics and Safety Problems **76**, 665–673 (2013)
37. Mahgrefteh, H., Brown, S., Zhang, P.: A dynamic boundary ductile-fracture-propagation model for CO_2 pipelines. Journal of Pipeline Engineering **9**(4) (2010)
38. Spinelli, C.M., Demofonti, G., Di, B.M., Lucci, A.: CO_2 full scale facilities challenges for EOR/CCTS testing on transportation issues. In: The Twenty-Second International Offshore and Polar Engineering Conference, Rhodes, Greece, ISOPE-I-12-323. OnePetro (2012)
39. Koornneef, J., Ramirez, A., Turkenburg, W., Faaij, A.: The environmental impact and risk assessment of CO_2 capture, transport and storage–an evaluation of the knowledge base. Prog. Energ. Combust. **38**(1), 62–86 (2012)
40. Veritas, D.N.: Design and operation of CO_2 pipelines, Norway, Det Norske Veritas, 1–42 (2010)

Chapter 4
Corrosion in CO_2 Geological Utilization and Storage

Yongcun Feng, Wei Yan, Liwei Zhang, and Yan Wang

4.1 General Introduction

Corrosion in CO_2 geological utilization and storage is a common phenomenon, and a widely-observed impact of such corrosion is the impairment of wellbore integrity. Wellbore integrity can be defined as the degree of consolidation of the casing-cement-formation rock system through the bonding of well cement, and the ability to maintain the mechanical strength and sealing capacity of the casing-cement-formation rock system [1, 2]. Wellbore integrity includes the integrity of cement ring and the integrity of interfaces (i.e., the casing-cement ring interface, and the cement ring-formation rock interface). In engineering practices, both the wellbore casing and the wellbore cement are prone to integrity failure, and the main failure modes are divided into the failure caused by mechanical force and the failure caused by chemical corrosion.

For mechanical failure, both the steel casing and cement ring of the wellbore may fail under the action of external forces. Compared with steel casing and other downhole metal components, wellbore cement ring, as a porous medium containing natural defects, is more prone to damage and failure under complex loading conditions downhole. The wellbore cement ring not only provides mechanical support for the casing, but also plays a key role in interlayer separation and leakage prevention.

Y. Feng (✉)
College of Petroleum Engineering, China University of Petroleum (Beijing), Beijing 102249, China
e-mail: yfeng@cup.edu.cn

L. Zhang · Y. Wang
State Key Laboratory of Geomechanics and Geotechnical Engineering, Institute of Rock and Soil Mechanics, Chinese Academy of Sciences, Wuhan 430071, China

University of Chinese Academy of Sciences, Beijing 100049, China

W. Yan
Unconventional Petroleum Research Institute, China University of Petroleum (Beijing), Beijing 102249, China

L. Zhang (ed.), *Corrosion in CO_2 Capture, Transportation, Geological Utilization and Storage*, Engineering Materials, https://doi.org/10.1007/978-981-99-2392-2_4

The mechanical failure of the cement ring will lead to the loss of effective wellbore barriers, resulting in gas leakage and an increase in the annulus pressure at the casing-cement ring and the cement ring-formation rock interfaces, affecting safe CO_2 injection operations, and even causing severe accidents such as blowouts.

For chemical failure, both the steel casing and the cement ring of the wellbore can be chemically corroded under the influence of CO_2 and other corrosive agents. For steel casing, its corrosion can be classified as internal corrosion and external corrosion. External corrosion is mainly formation fluid-induced corrosion at the exterior of the casing, combined with stray current-induced corrosion. Internal corrosion is mainly caused by the internal fluid inside the casing, usually caused by an intrusion of formation fluid or CO_2 into the annulus space between the casing and the oil tubing. In the CCUS environment, the corrosion of steel casing has three characteristics: (1) Multiphase flow corrosion in which gas, water, and solid particles (or hydrocarbons if the project is a CO_2-enhanced oil recovery project) coexist; (2) High pressure and/or high temperature environment; (3) Corrosion is usually the result of the combined action of CO_2, H_2S, O_2, Cl–, SO_4^{2-}, moisture, etc.. For cement rings, due to the fact that the ordinary wellbore cement mainly contains tricalcium silicate ($3CaO{\cdot}SiO_2$), dicalcium silicate ($2CaO{\cdot}SiO_2$), tricalcium aluminate ($3CaO{\cdot}A_2O_3$), and tetracalcium ferroaluminate ($4CaO{\cdot}A_2O_3{\cdot}Fe_2O_3$) and they produce large amounts of $Ca(OH)_2$ during hydration, a significant decrease in pH due to the dissolution of CO_2 in the formation fluid can cause fast dissolution of alkaline $Ca(OH)_2$ in cement. This chapter discusses the wellbore failure caused by geomechanical force (Sect. 4.2) and the wellbore failure caused by chemical corrosion (Sect. 4.3) in detail.

4.2 Wellbore Failure Caused by Geomechanical Force

4.2.1 Failure Mode

Similar to concrete, cement rings in CCUS wells can withstand large compressive stress, but their ability to withstand tensile stress is weak. The tensile strength of the cement ring is much smaller than its compressive strength (about 15 ~ 20% of its compressive strength). Due to the heterogeneous nature of cement, the cement ring is with a large number of natural defects, and the internal microscopic pores in cement is prone to stress concentration, which causes the cement ring to show obvious brittleness, and it is easy to deform and fail under the action of external force. In addition, typical wellbore cement is accompanied by significant hydration shrinkage during the solidification and hardening process (the shrinking volume is usually 4 ~ 6% of the total volume), resulting in micro-gaps at the cementing interface between the cement ring and formation rock. Given the characteristics of the cement ring of CCUS wells, under the complex loading conditions of downhole, the cement ring of CCUS wells has the following typical failure modes: (1) casing-cement ring interface separation. When the casing and the cement ring are deformed incongruously, there

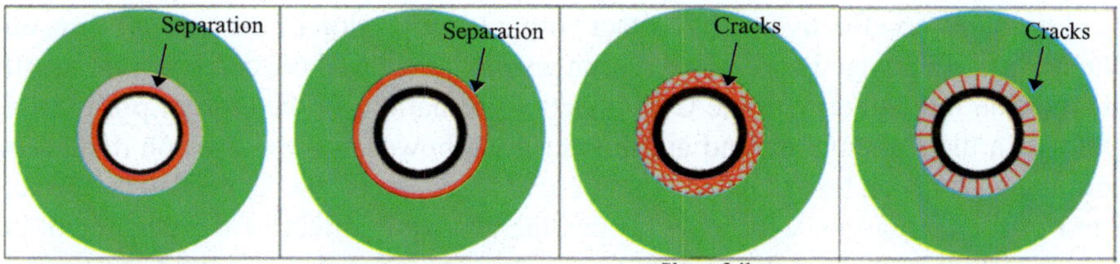

Fig. 4.1 Schematic of typical failure modes of the cement ring [3]

is a risk of cementing interface separation. For example, when the low-temperature CO_2 is injected into the downhole, due to the different thermodynamic properties of the casing and the cement ring, the degrees of cold shrinkage of the casing and the cement ring are different, resulting in micro-gaps at the casing-cement ring interface. (2) Cement ring-formation rock interface separation. The cement hardening process of CCUS wells is usually accompanied by hydration contraction, resulting in micro-gaps at the cement ring-formation rock interface. (3) Shear failure. When the cement ring is under a large deflection stress, shear failure is prone to occur. (4) Radial cracks. When the circumferential stress of the cement ring exceeds the tensile strength, the cement ring will have tensile failure. When the pressure on the inner wall of the cement ring is greater than the pressure on the outer wall of the cement ring, and there is a contraction of the cement ring, it is easy to form radial cracks in the cement ring (Fig. 4.1).

4.2.2 Influencing Factors of Wellbore Failure

Under CO_2 geological utilization and storage conditions, the factors affecting the potential of wellbore failure include geological conditions, the quality of drilling and completion engineering, and the material properties of the cement ring. In order to maintain the long-term integrity of the cement ring of CCUS wells, it is necessary to analyze the influencing factors and understand the impacts of the factors.

4.2.2.1 Geological Conditions

The influence of acid fluid. CCUS projects inject a large amount of CO_2 into the CO_2 storage reservoirs. CO_2 produces carbonic acid (H_2CO_3) in the water phase, which reduces the pH of the formation water and causes corrosion to the cement ring of the CCUS wells. The corrosion may lead to cracking and decalcification of the cement ring, which damages the structure of the cement ring and aggravates the failure risk of the CCUS well under the complex loading conditions of the downhole. In the case of no pre-existing defects in the cement ring and at the interfaces, the acid fluid has no effective path to react with a large portion of the cement ring, and the corrosion damage is restricted within a small portion of the cement ring, mostly the portion in contact with the CO_2 injection formation. In the case of pre-existing defects in the cement ring and at the interfaces, however, the corrosion damage in the cement ring can be extensive. The pre-existing defects can be invaded by the acid fluid of the formation, which aggravates the corrosion effect of the acid fluid and causes a large portion of the cement ring to be exposed to the acid fluid. As a result, a large length of the cement ring can be damaged in vertical direction, and the failure risk of the CCUS well is greatly increased.

The influence of high pressure and high temperature conditions. CO_2 storage formations usually have high pressure, which requires a high drilling fluid density and a high solid phase content of drilling fluid. Such drilling fluid is easy to form thick and loosely cemented filter cake on the wall of the well, and it is difficult to clean the filter cake during the well cementing process, resulting in poor bonding quality of the cement-formation rock interface. Also, during the cement hardening process of the CCUS well, the hydrostatic pressure of the annular cement slurry column gradually decreases. When the hydrostatic pressure of the cement slurry column is lower than the formation pressure, the formation fluid is easy to invade the space occupied by cement slurry, forming fluid channels in the cement ring that is difficult to repair. Formation fluid can travel upward via the fluid channels or the cementing interfaces to the ground, which causes a failure of the cement ring integrity.

Under high temperature conditions, the decline of cement ring strength of CCUS wells is possible. The C–S–H gel in wellbore cement formed as a result of cement hydration reaction can exist in a stable state at room temperature, and the C–S–H gel plays a key role in the strength and structural stability of the cement ring. When the temperature is higher than 110 °C in the formation, the C–S–H gel is no longer stable and is converted to hydrated dicalcium silicate with low compressive strength and high permeability. Under high temperature conditions, the conversion of C–S–H gel into hydrated dicalcium silicate in the cement ring causes the structure of the cement ring to change, which results in a reduction of the strength and an increase of the permeability [4]. When pure Portland cement was cured at 230 °C for one month, its compressive strength dropped to one-third of the original compressive strength. Also, the permeability became 10 ~ 100 times of the original permeability, and the integrity of the cement was impaired [5].

The influence of formation heterogeneity. Formation heterogeneity represents the spatial distribution of different strata and varying rock properties in formation layers. Formation heterogeneity is attributed to sedimentation, diagenesis and late tectonic action in the long geological history. Fractures, fragments and anisotropy are typical features of formation heterogeneity. During the sedimentation process, the rock is constantly subjected to complex geological actions, resulting in different internal structures such as strata, joints, voids, holes and uneven distribution of rock particles. Therefore, formation heterogeneity is very common. Anisotropy is the basic property of most natural rock materials, contributing to different properties in rock mechanics and varying deformation characteristics in different directions. In formations with high heterogeneity, the in-situ stress on the cement ring is complicated, which exacerbates the integrity failure of the cement ring.

4.2.2.2 The Quality of Drilling and Completion Engineering

Drilling fluid residue. During the drilling process, the drilling fluid forms an adhesion layer at the casing-cement and cement-formation interfaces. Under normal circumstances, it is difficult to remove the adhesion layer attached to the casing string and the formation rock in the cementing process. Due to the hydrophobic nature of the drilling fluid residue, the hydrophilic cement slurry cannot create a good bonding at the interfaces with the drilling fluid residue. As a result, locations along the interfaces with drilling fluid residue become the weak points of the cementing interfaces, and gas channeling is prone to occur at those locations under the complex loading conditions of the downhole.

Eccentric casing. During drilling at a great depth, it is difficult to build a regular borehole structure, and the casing string is usually not exactly centered in the borehole. In actual engineering, casing eccentricity affects the drilling fluid flushing efficiency in cementing engineering. The flow rate of the fluid in the annular space is negatively correlated with the cross-sectional area of the annular space. As a result, the flow rate of the narrow-side fluid in the annular space given an eccentric casing is higher than that of the wide-side fluid, so drilling fluid retention at the wide side in the annular space is likely to occur during the process of flushing and cement injection due to a low flow rate, resulting in bad cementing job. At the same time, the eccentricity of the casing causes an uneven thickness of the cement ring, which in turn leads to uneven distribution of the stress state in the cement ring under the action of formation pressure loading and injection pressure loading. The uneven distribution of the stress state in the cement ring affects the long-term integrity of the cement ring.

Pressure in the well. Casing strength test, fluid injection, fluid extraction, etc. cause pressure changes in the well. If the pressure in the well becomes high, the cement ring of the well enters plastic deformation stage from elastic deformation and is prone to damage. On the other hand, if the pressure in the well becomes low, the wellbore shrinks inward, and due to the difference in mechanical properties, the casing and the cement ring have uncoordinated deformation, resulting in a micro-gap at the interface between casing and cement ring.

Perforation operations. During perforated completion of CCUS wells, the water jets produced by shaped perforation projectiles penetrate through the casing wall, cement rings, and parts of the formation of CCUS wells. Besides water jets, bullet gun and shaped charges are also common perforating methods used to initiate a hole from the wellbore through the casing and any cement sheath into the producing zone. After perforation, some holes along the perforated spiral appear on the cement ring, resulting in a decrease in the structural stability of the cement ring. In the subsequent production process, due to excessive load, the cement ring with perforated holes is likely to have structural failure such as crushing or pulling, which has a non-negligible impact on the structural integrity of the perforated section of cement.

4.2.2.3 The Material Properties of the Cement Ring

The mechanical properties of the cement ring have a direct impact on its integrity. As mentioned above, the cement ring has a poor tensile strength, and it can be easily damaged under the action of external forces. By enhancing the deformation ability of the cement ring under external force, the failure risk of the cement ring can be reduced. That is to say, a flexible cement ring with low elastic modulus and high Poisson's ratio is conducive to maintaining the long-term integrity [6]. In actual engineering, some parameters of the cement ring need to be appropriately designed to make the cement ring compatible with the formation rock and the casing. For example, the thickness of the cement ring cannot be too large or too small, and the optimal thickness needs to be determined according to the diameter of the borehole, the pressure of the formation, and the diameter of the casing [7–9].

4.3 Mechanical Strength Analysis for Wellbore Integrity Evaluation

The mechanical strength analysis for wellbore integrity evaluation includes mechanical strength analysis on cement ring matrix, and mechanical strength analysis on the interfaces of casing-cement and cement-formation rock. The mechanical strength analysis is the fundamental basis for the analysis of corrosion-induced mechanical strength failure. The integrity of the cement ring is closely related to the properties of the formation rock and the casing, which are in direct contact with the cement ring. Since Zinkham and Goodwin proposed the concept of "the assembly of the casing, the cement ring, and the formation rock" (hereafter referred to as "the assembly") in 1962 [10], scholars have taken the assembly as the main research object of wellbore integrity, and scholars have proposed a variety of analysis methods to study the integrity of the assembly. In this section, based on the traditional elastic mechanical method, the stress distribution of the assembly of the casing, the cement ring, and the formation rock is studied and analyzed, and the influence of the cement ring performance on the stress state of the wellbore assembly is discussed.

To better explain the methodology, the influence of cementing quality on the integrity of the cement ring is ignored and it is assumed that: (1) the casing and the cement ring are intact and defect-free; (2) The bonding of the two cementing interfaces is good, and there is no relative slippage; (3) The casing is completely centered; (4) The casing, cement ring and formation are all uniform isotropic bodies; (5) The cement ring has no initial stress; (6) Uniform distribution of in-situ stress. Given

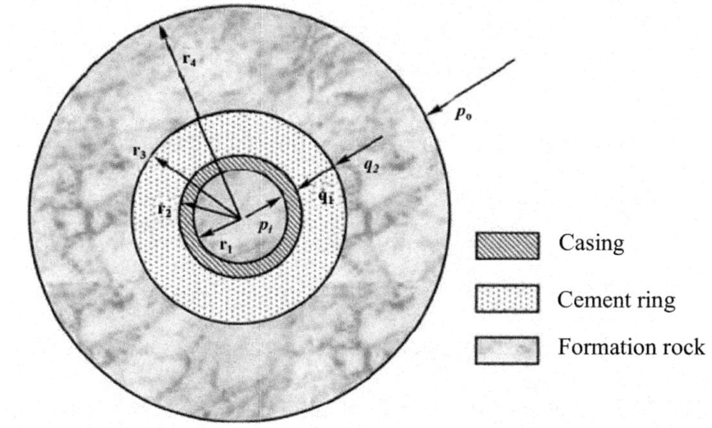

Fig. 4.2 A schematic of the casing-cement ring-formation assembly

the aforementioned assumptions, the problem of the casing-cement ring-formation assembly is transformed into a two-dimensional plane strain problem, as shown in Fig. 4.2.

(1) Casing

Based on Lame equation, the stress distribution inside a casing can be expressed as:

$$\begin{cases} \sigma_r = \frac{r_1^2 r_2^2 (q_1 - p_i)}{r_2^2 - r_1^2} \frac{1}{r^2} + \frac{r_1^2 p_i - r_2^2 q_1}{r_2^2 - r_1^2} \\ \sigma_\theta = -\frac{r_1^2 r_2^2 (q_1 - p_i)}{r_2^2 - r_1^2} \frac{1}{r^2} + \frac{r_1^2 p_i - r_2^2 q_1}{r_2^2 - r_1^2} \end{cases} \quad r_1 \leqq r \leqq r_2 \tag{4.1}$$

The radial displacement of the casing wall can be calculated as:

$$\begin{aligned} u_{so} &= \frac{1 + \upsilon_s}{E_s} \left[-\frac{r_1^2 r_2^2 (q_1 - p_i)}{r_2^2 - r_1^2} \frac{1}{r_2} + \frac{r_1^2 p_i - r_2^2 q_1}{r_2^2 - r_1^2} r_2 \right] \\ &= \frac{1 + \upsilon_s}{E_s} \frac{2(1 - \upsilon_s) r_1^2 r_2}{r_2^2 - r_1^2} p_i - \frac{1 + \upsilon_s}{E_s} \frac{r_1^2 r_2 + (1 - 2\upsilon_s) r_2^3}{r_2^2 - r_1^2} q_1 \\ &= f_1 p_i - f_2 q_1 \end{aligned} \tag{4.2}$$

In Eq. (4.2), E_s is the elastic modulus of the casing, and υ_s is the Poisson's ratio of the casing.

(2) Cement ring

The stress distribution in the cement ring can be expressed as:

$$\begin{cases} \sigma_r = \frac{r_2^2 r_3^2 (q_2 - q_1)}{r_3^2 - r_2^2}\frac{1}{r^2} + \frac{r_2^2 q_1 - r_3^2 q_2}{r_3^2 - r_2^2} \\ \sigma_\theta = -\frac{r_2^2 r_3^2 (q_2 - q_1)}{r_3^2 - r_2^2}\frac{1}{r^2} + \frac{r_2^2 q_1 - r_3^2 q_2}{r_3^2 - r_2^2} \end{cases} \quad r_2 \leqq r \leqq r_3 \tag{4.3}$$

The radial displacement of the inner wall of the cement ring can be expressed as:

$$\begin{aligned} u_{ci} &= \frac{1+\upsilon_c}{E_c}\frac{r_3^2 r_2 + (1-2\upsilon_c) r_2^3}{r_3^2 - r_2^2} q_1 - \frac{1+\upsilon_c}{E_c}\frac{2(1-\upsilon_c) r_3^2 r_2}{r_3^2 - r_2^2} q_2 \\ &= f_3 q_1 - f_4 q_2 \end{aligned} \tag{4.4}$$

In Eq. (4.4), E_c is the elastic modulus of the cement ring, and υ_C is the Poisson's ratio of the cement ring.

The radial displacement of the exterior wall of the cement ring can be expressed as:

$$\begin{aligned} u_{co} &= \frac{1+\upsilon_c}{E_c}\frac{2(1-\upsilon_c) r_2^2 r_3}{r_3^2 - r_2^2} q_1 - \frac{1+\upsilon_c}{E_c}\frac{r_2^2 r_3 + (1-2\upsilon_c) r_3^3}{r_3^2 - r_2^2} q_2 \\ &= f_5 q_1 - f_6 q_2 \end{aligned} \tag{4.5}$$

(3) Formation

The stress distribution in the cement ring can be expressed as:

$$\begin{cases} \sigma_r = \frac{r_3^2 r_4^2 (p_o - q_2)}{r_4^2 - r_3^2}\frac{1}{r^2} + \frac{r_3^2 q_2 - r_4^2 p_o}{r_4^2 - r_3^2} \\ \sigma_\theta = -\frac{r_3^2 r_4^2 (p_o - q_2)}{r_4^2 - r_3^2}\frac{1}{r^2} + \frac{r_3^2 q_2 - r_4^2 p_o}{r_4^2 - r_3^2} \end{cases} \quad r_3 \leqq r \leqq r_4 \tag{4.6}$$

The radial displacement of the formation in contact with the cement ring can be expressed as:

$$\begin{aligned} u_{fi} &= \frac{1+\upsilon_f}{E_f}\frac{r_4^2 r_3 + (1-2\upsilon_f) r_3^3}{r_4^2 - r_3^2} q_2 - \frac{1+\upsilon_f}{E_f}\frac{2(1-\upsilon_f) r_4^2 r_3}{r_4^2 - r_3^2} p_o \\ &= f_7 q_2 - f_8 p_o \end{aligned} \tag{4.7}$$

In Eq. (4.7), E_f is the elastic modulus of the formation, and υ_f is the Poisson's ratio of the formation.

Because there is no relative sliding between casing, cement ring and formation, and the casing, the cement ring and the formation are closely connected, so the radial pressure between the components is the same and the radial displacement is continuous. The continuous conditions according to the interface displacement are:

$$\begin{cases} u_{so} = u_{ci} \\ u_{co} = u_{fi} \end{cases} \tag{4.8}$$

A combination of (4.2), (4.4), (4.5), (4.7) and (4.8) yields:

$$\begin{cases} q_1 = \frac{f_4 f_8 p_o + (f_6 + f_7) f_1 p_i}{(f_2 + f_3)(f_6 + f_7) - f_4 f_5} \\ q_2 = \frac{(f_2 + f_3) f_8 p_o + f_1 f_5 p_i}{(f_2 + f_3)(f_6 + f_7) - f_4 f_5} \end{cases} \quad (4.9)$$

In Eq. (4.9), q_1 is the radial pressure at the casing-cement ring interface, and q_2 is the radial pressure at the cement ring-formation interface.

According to the aforementioned stress analysis method of the casing-cement ring-formation assembly, as long as the elastic parameters of the wellbore, geometric dimensions of the wellbore, pressure distribution in the wellbore, and the stress field in the formation are given, the stress distribution of the casing-cement ring-formation assembly can be determined. However, it is still challenging to combine corrosion with mechanical strength analysis, primarily due to the complex nature of corrosion reactions and the difficulty in obtaining justified mechanical strength parameters for each corrosion product. CO_2-induced corrosion turns the casing-cement ring-formation assembly into a highly heterogeneous system with corrosion products accumulated at certain locations within the assembly, and the stress analysis on such a system is complicated.

4.4 Wellbore Failure Caused by Chemical Corrosion

4.4.1 Steel Corrosion

The multiphase flow in CO_2 geological utilization and storage scenarios involves gas, liquid and solid phases (mainly sand grains). Due to the interaction between the phases, the multiphase system is more corrosive than a single-phase system. For steel casings exposed to the multiphase system, they may experience severe corrosion. In addition, some CGUS projects have high temperature and high pressure environments, and the corrosion laws and mechanisms of the metals in high temperature and high pressure environments are often different from the cases in room temperature and pressure.

Given large amounts of CO_2 injection into the subsurface for the purpose of geologic CO_2 utilization and storage, CO_2 corrosion is expected to cause degradation of subsurface infrastructure in geologic CO_2 utilization and storage projects, especially in wellbore steel casing. Under geologic CO_2 utilization and storage scenarios, part of the injected CO_2 is dissolved in the formation water, and carbonic acid is formed when dissolved CO_2 reacts with water [11, 12]. Corrosion occurs when steel comes in contact with carbonic acid. Although carbonic acid is a weak acid, it may corrode steel at a rate higher than that of hydrochloric acid under the same pH [13]. If CO_2 is injected into a formation with limited or no presence of pH-buffering minerals, the formation water pH can drop to < 3 with the abundance of injected CO_2 [14]. Therefore, the environment of geologic CO_2 utilization and storage is suitable for

Fig. 4.3 Cracks and pits formed on steel casings due to CO_2 corrosion [15]

accelerated steel corrosion. CO_2 corrosion in the aggressive downhole environment can cause severe steel casing damage (Fig. 4.3). To prevent accidents and economic losses caused by CO_2-induced steel corrosion, steel corrosion evaluation and control is a must for geologic CO_2 utilization and storage projects.

The steel of the casing is a good conductor, and when the steel comes into contact with formation water rich in CO_2 and dissolved ions, the steel forms a circuit as a good conductor of electricity, and iron in steel tends to lose electrons to produce positively charged ferrous ions (Fe^{2+}), which then dissolves in the formation water. Electrons tend to congregate at the metal end, forming a certain potential difference, and then enter the formation water as well. The electrons entering the formation water are bounded by H^+ in water to produce H_2. In an aerobic environment, OH^- instead of H_2 is produced. In a CO_2-rich environment, Fe^{2+} reacts with $HCO_3{}^-$ to form $FeCO_3$. In summary, The CO_2-induced steel corrosion process involves the following steps:

(1) Dissolution of CO_2 in water to form H_2CO_3:

$$CO_2 + H_2O \rightarrow H_2CO_3 \tag{4.10}$$

(2) Disassociation of H_2CO_3 to produce H^+:

$$H_2CO_3 \rightarrow H^+ + HCO_3^- \tag{4.11}$$

(3) Iron dissolution to release electrons:

$$Fe\ (s) \rightarrow Fe^{2+} + 2e^- \tag{4.12}$$

(4) H^+ accepts electrons to form H_2:

$$2H^+ + 2e^- \rightarrow H_2(g) \tag{4.13}$$

(5) Under oxidative conditions, OH^- instead of H_2 (g) is produced:

$$2H_2O + O_2(g) + 4e^- \rightarrow 4OH^- \quad (4.14)$$

(6) Fe^{2+} in solution reacts with $HCO_3{}^-$ to form $FeCO_3$:

$$Fe^{2+} + HCO_3^- \rightarrow FeCO_3(s) + H^+ \quad (4.15)$$

(7) Under oxidative conditions, $FeCO_3$ can be oxidized by O_2 to form Fe_2O_3, Fe_3O_4 or $Fe(OH)_3$:

$$4FeCO_3(s) + O_2(g) + 4H_2O \rightarrow 2Fe_2O_3(s) + 4HCO_3^- + 4H^+ \quad (4.16)$$

$$6FeCO_3(s) + O_2(g) + 6H_2O \rightarrow 2Fe_3O_4(s) + 6HCO_3^- + 6H^+ \quad (4.17)$$

$$4FeCO_3(s) + O_2(g) + 10H_2O \rightarrow 4Fe(OH)_3(s) + 4HCO_3^- + 4H^+ \quad (4.18)$$

(8) If the environment changes from acidic condition to alkaline condition, $FeCO_3$ undergoes dissolution and soluble Fe (II)-carbonate complex is created:

$$FeCO_3(s) + HCO_3^- \rightarrow Fe(CO_3)_2^{2-} + H^+ \quad (4.19)$$

There are two different reaction mechanisms to explain the role of carbonate ions on cathodic reactions: "buffering" mechanism and "direct reduction" mechanism. The two mechanisms are not mutually exclusive, but it is crucial to clarify which mechanism plays a vital role in a given environment, so as to better predict the rate of CO_2 corrosion. The buffering mechanism only considers the reduction of H^+ ions, and H_2CO_3 and $HCO_3{}^-$ only act as "agents" for storing H^+ ions. In the direct reduction mechanism, positive monovalent hydrogens in H_2CO_3 and $HCO_3{}^-$ can be directly reduced [16–18]. The direct reduction mechanism is often used to explain the phenomenon of higher corrosion rates when CO_2 is dissolved in aqueous solutions with high concentrations. Nesic et al. [19] noted that the corrosion limit current in aqueous solutions containing CO_2 was higher than that in HCl solutions at the same pH, and they attributed that phenomenon to direct reduction of H_2CO_3 and $HCO_3{}^-$ on metal surfaces. Gulbransen and Bilkova [20] conducted linear polarization resistance (LPR) and weight loss measurements to study the process of metal corrosion by CO_2, and they found that the direct reduction of carbonic acid and acetic acid played very important roles in determining the overall corrosion rates.

Steel pipelines and valves at the surface or close to the surface are prone to reactions 4.16, 4.17 and 4.18 due to abundant presence of O_2. For well casings at deep subsurface with the presence of high concentration CO_2, both the O_2 concentration and the pH are very low, and the main corrosion product is $FeCO_3$.

$FeCO_3$ tends to deposit on the casing surface and has an effect on delaying further corrosion of the casing. The SEM images at the top of a CO_2-corroded carbon steel

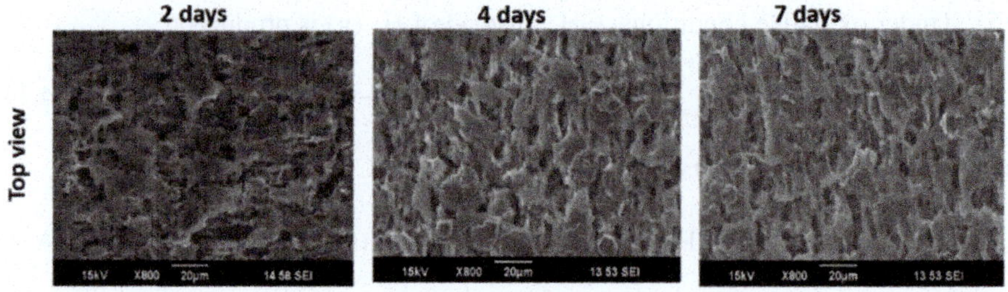

Fig. 4.4 SEM images of the top view of a CO_2-corroded carbon steel [21]

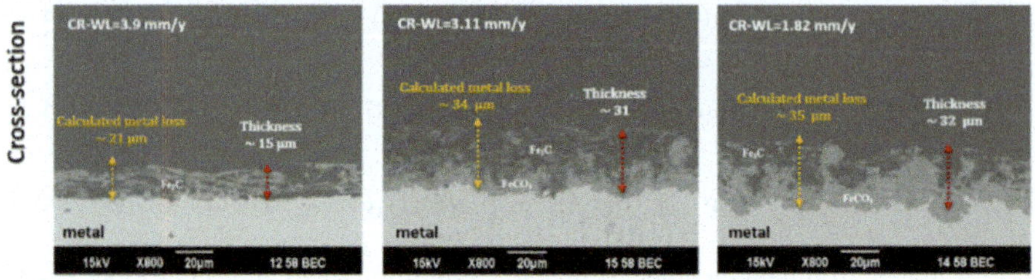

Fig. 4.5 SEM images of the cross-section view of a CO_2-corroded carbon steel [21]

reveal that the $FeCO_3$ layer becomes denser with the increase of corrosion time (Fig. 4.4). A cross-section view of the same steel shows the growth of the $FeCO_3$ layer and a thin, porous iron carbide (Fe_3C) layer outside the $FeCO_3$ layer, which is the left-over component of the original steel after iron gets dissolved (Fig. 4.5).

Whether the $FeCO_3$ deposition layer can effectively delay casing corrosion depends on the size and arrangement of the $FeCO_3$ deposition grains. If a dense and non-porous depositionlayer forms, it can cover the steel surface and acts as a diffusion barrier between the corrosive species, such as H^+, and the metal surface [22]. Even a thin layer (4–6 μm thick) of a dense $FeCO_3$ deposition can offer a good protectiveness and reduce corrosion rate. However, if the $FeCO_3$ precipitation rate is lower than the corrosion rate, a porous and nonprotective $FeCO_3$ scale will form, which cannot inhibit steel corrosion [22].

CO_2 corrosion of steel has two types: uniform corrosion and local corrosion. When uniform corrosion occurs, the partial reactions (iron dissolution and H^+ reduction) are statistically distributed over the steel surface, leading to relatively homogeneous dissolution of the steel and uniform formation of corrosion products. In short, the casing is evenly damaged in all or most of the area in contact with CO_2. As for local corrosion, some parts of the metal surface are severely corroded, while others are less corroded. Local corrosion is caused by electrochemical inequality, which results in the formation of local batteries. Local corrosion can be quite problematic. Whereas uniform corrosion can be seen clearly on the surface, local corrosion often appears only as small pinholes on the surface. The amount of iron removed below the pinholes is generally unknown (hidden cavities can form below the pinholes), making local corrosion more difficult to detect and predict. Furthermore, local corrosion may lead to stress concentration, which then causes stress-induced cracking. Stress-induced cracking due to corrosion is a combined mechanical and electro-chemical process that results in cracking of steel. Stress-induced cracking may cause unexpected sudden brittle failure of normally ductile steel. In summary, the impact of local corrosion is generally higher than that of uniform corrosion.

The corrosion area is expected to expand with the increase of corrosion time. An expansion of corrosion area can cause corrosion products to accumulate and connect together to form a film structure. Depending on the steel reinforcement material and corrosion environment, either a single-layer or a multi-layer corrosion product film is deposited on the surface of steel reinforcement. The processes of corrosion product film formation and growth are summarized in Fig. 4.6. As previously discussed, it is difficult for external corrosion environment to directly contact the surface of steel reinforcement with the improvement of a film structure [23]. Therefore, the rapid formation of corrosion product film with good integrity can be helpful to weaken the structural deterioration of steel reinforcement after CO_2 corrosion.

Overall, steel corrosion of wellbores under CCUS conditions involves almost all types of corrosion, which can be summarized as follows:

(1) Weak acid corrosion. CO_2, H_2S, etc. dissolve in water to form weak acid, resulting in corrosion.

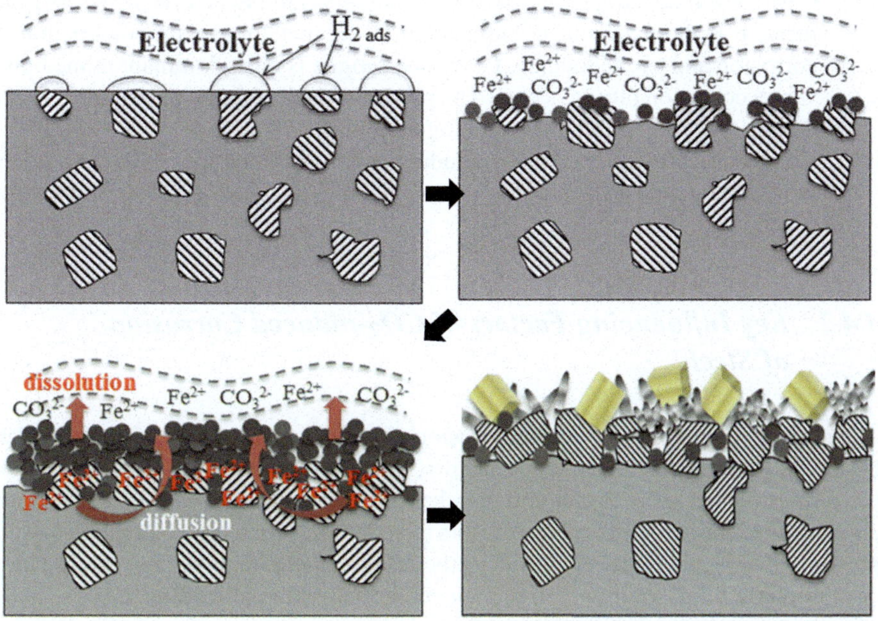

Fig. 4.6 Schematic diagrams showing the formation and growth of the corrosion product film (modified from Li et al. [24])

(2) Contact corrosion. Contact corrosion refers to the electrochemical corrosion that occurs when different materials are in contact with each other in the electrolyte solution. Contact materials include metallic and non-metallic materials, and the contact corrosion between different metals is usually called galvanic corrosion.
(3) Crevice corrosion. Crevice corrosion refers to the corrosion that occurs at small gaps and joints, e.g., at the joints of oil tubes and casings.
(4) Pitting corrosion. Pitting corrosion refers to the corrosion forming small holes due to abrasions, defects, coverage by mud and dust, etc.
(5) Stray current corrosion. Stray current corrosion is mainly caused by stray alternating current (AC) or direct current (DC) outside the well casing. When the current intensity is the same, the corrosion caused by DC is 100 times greater than that of AC.
(6) Cavitation corrosion. Cavitation corrosion is mainly caused by the hammering effect of bubble rupture on the surface of the metal.
(7) Erosion-assisted corrosion. Erosion is a physical phenomenon on the surface of the metal caused by solid particle scraping (such as sand particles) in the flow under the action of fluid mechanics. Erosion can cause corrosion product film damage, thereby promoting corrosion.
(8) Flow-assisted corrosion. The flow leads to accelerated mass transfer rates of the reaction media and corrosion products, and accelerated reaction rates on metal surfaces, thereby raising the corrosion rate.
(9) Corrosion-induced fatigue. Corrosion fatigue is fatigue in a corrosive environment. It is the mechanical degradation of the metal under the joint action of corrosion and cyclic loading. Corrosion fatigue is most dominant in mediums where corrosion pitting occurs. The pits act as stress raisers and initiate fatigue cracks, which lead to corrosion fatigue failure.
(10) Corrosion-induced cracking, including hydrogen bubbling (HB), hydrogen-induced cracking (HIC), sulfide stress cracking (SCC), chloride stress cracking (CSC), etc.

4.4.2 Key Influencing Factors of CO_2-induced Corrosion of Steel

The main factors of CO_2-induced corrosion of steel casing include inherent properties of the steel and environmental factors. Inherent properties of the steel include the microstructure of the steel and the chemical compositions of the steel (mainly alloying elements). Environmental factors include water content, temperature, partial pressure of CO_2, chemical properties of the solution medium, flow rate, single-phase or multiphase fluid, solution pH, scaling conditions, external loading, etc.

4.4.2.1 Inherent Properties of the Steel

The influence of steel microstructure. From a metallographic point of view, the amount of carbonates deposited on steel surface (mainly $FeCO_3$) increase as CO_2 corrosion progresses. These carbonates form a porous, spongy, flaky or needle-like structure on the surface, which favors deposition of additional carbonates. The shape and structure of the carbonate deposition layer depend on the original metallographic structure of the steel. The $FeCO_3$ layer of the normalizing 45 steel sample is generally thicker than that of the quenched & tempered (Q&T) steel sample. At the same time, the adhesion and thicknesses of other reaction layers on steel surface also depend on the microstructure of the steel. For example, the formation of the passivation film on the surface of the normalizing 45 steel sample is faster than that of the Q&T steel sample.

The influence of alloying elements. Alloying elements have a great influence on CO_2-induced steel corrosion. For example, when a small amount of Cu is added to the steel, the activation energy of CO_2 hydrolysis to generate carbonic acid is greatly reduced, thereby raising the rate of carbonic acid formation and accelerating the corrosion rate. Improving the Cr content in the steel matrix can decrease the uniform corrosion rate [25]. Specifically, if there are a large amount of chromium in steel reinforcement and a large amount of sulfur in the environment, multiple layers of corrosion products are accumulated [26]. In that scenario, the films with corrosion products are composed of chromium hydroxide, chromium oxide, ferrous carbonate, and iron-sulfur compounds, and the films are denser and less permeable compared with $FeCO_3$-only films [27]. The study by Koguma et al. pointed out that the addition of chromium and molybdenum to the steel raised the resistance of the steel to CO_2 corrosion [28]. Sun et al. [29] conducted corrosion tests on steels with different Cr contents, and they found that coarse grains of $FeCO_3$ were formed on the surface of carbon steel and low chromium (less than 1 wt%) steel. After removing the corrosion product film on the surface, severe localized corrosion pits in steel with no or very low Cr content were observed. Therefore, the CO_2 corrosion resistance of carbon steel and low chromium steel is low [29]. Different from carbon steel and low chromium steel, 13Cr steel, 25Cr steel, and α-γ duplex stainless steel are resistant to CO_2 corrosion. In several cases, the steel with a 2 wt% Cr content can effectively reduce the corrosion rate. The corrosion resistance of Cr-rich steel is mainly due to the enrichment of Cr elements in the corrosion product film. For example, in steel with a Cr content of 2 wt%, the Cr concentration in the corrosion product film can reach 15–17%, which is 7.5–8.5 times of the Cr content in steel. In a humid environment, the corrosion product film of the Cr-bearing steel is dense and has good adhesion and toughness.

4.4.2.2 Influence of Environmental Factors

The influence of water content. The presence of H_2O is a prerequisite for steel corrosion in a supercritical CO_2 environment. Supercritical CO_2 with no water presence is not corrosive to steel casing. However, the presence of water in deep subsurface formations always introduces water vapor into the injected supercritical CO_2, and the supercritical CO_2 with water presence (above the critical water content) becomes very corrosive to steel casing. A comparison between corrosion induced by supercritical CO_2 with water vapor and dissolved CO_2 shows that the level of corrosion induced by supercritical CO_2 is more severe than that induced by dissolved CO_2. In dissolved CO_2, the corrosion layer on the surface of the steel is uniform and the thickness of the corrosion layer is uniform as well. However, in supercritical CO_2 with water vapor, pitting corrosion is obvious, and localized and deep damages are common.

Field experience and experimental results show that as long as the water content in supercritical CO_2 is lower than the critical water content, the corrosion rate of the steel will be small, regardless of the presence or absence of impurities. Compared to reducing impurities, reducing the water content is the most effective way to prevent corrosion [30]. For CO_2 transport pipelines, it is relatively easy to control the water content. For oiltubes and steel casings of wellbores, since formation water is present in most scenarios, the oiltubes and steel casings of wellbores are usually exposed to a water-rich environment, so the corrosion of oiltubes and steel casings of wellbores is more severe than that of the CO_2 transport pipelines.

The impact of water content on the rate of steel corrosion in a supercritical CO_2 environment has been the subject of numerous studies. The two representations of the moisture degree are the actual water content (typically expressed in ppmv) and the critical relative humidity RH, which is defined as the ratio of the actual water content to the water solubility limit [31]. In order to investigate the corrosion behavior of carbon steel at 75.8 bar and 40 °C in a supercritical CO_2 environment with various water contents, Ayello et al. [32] used a thermodynamic model and electrochemical measurement technique. According to their electrochemical measurements, the corrosion rate rose as the water content rose, reaching 1.2 mm/year when the water content reached 100 ppm. In a supercritical CO_2 environment with O_2 and SO_2 at 80 bar and 50 °C, Choi and Nesic [33] examined the impact of water content on the corrosion rate of X65 steel. The findings demonstrated that the corrosion rate of X65 steel reached 1 mm/year when the water was saturated in the supercritical CO_2 stream (3300 ppm). When the amount of water was lower than 650 ppm, no corrosion phenomenon was observed.

In a crude oil/ScCO_2/brine multiphase environment, the water content also affects the state of the mixture, which in turn, affects the corrosion rate. In a supercritical CO_2 environment with crude oil, CO_2, and brine at 8 MPa, 90 °C, and 12 MPa, 65 °C, respectively, Farelas et al. [34] investigated the effects of water content on the corrosion of carbon steel. The findings demonstrated that the combination was in a water-in-oil state when the water percentage was less than 50%, and the corrosion rate essentially remained constant as the water content varied. But as the water concentration exceeded 50%, the combination changed into an oil-in-water state, and as the water content climbed, so did the average corrosion rate and pitting rate.

If there are impurities such as H_2S and SO_2 in the supercritical CO_2 environment, the critical water content that causes noticeable corrosion will decrease. Many scholars have studied the critical water content in various supercritical CO_2 environments with different impurities [35]. For example, Hua et al. [30] found that when the SO_2 concentration increased from 50 to 100 ppm under $ScCO_2$ environment with 20 ppm O_2, the critical water content changed from 2120 to 1850 ppmv.

The influence of temperature. Temperature, as one of the most important influencing factors in CO_2 corrosion, plays a vital role in the possible formation of corrosion product layer. The influence of temperature on CO_2 corrosion can be explained by several mechanisms. First, the temperature affects the solubility of CO_2. The concentration of CO_2 in the water decreases with an increasing temperature. Second, the temperature affects the mechanism of film formation of corrosion products. Third, the temperature directly determines the rates of corrosion reactions. Increasing temperature can change the chemical compositions of corrosion product films [36] and affect the CO_2 corrosion rate [37–39]. Because the corrosion product FeCO3 is relatively soluble at low temperatures, it is commonly accepted that at temperatures below 60 °C, no protective layer forms on the surface of steel [24]. Due to a decrease in solubility as temperature rises, $FeCO_3$ precipitates from the solution and forms dense and adherent corrosion product layers on the steel surface [40–42]. Therefore, a high temperature favors precipitation of $FeCO_3$, which hinders corrosion of steel. At high temperatures, the carbonic acid may interact with the steel, resulting in the creation of dense $FeCO_3$, which can effectively prevent steel corrosion. Therefore, at high temperatures, the disassociation of carbonic acid into bicarbonate and H^+ ions does not cause severe corrosion in steel. Therefore, corrosion rate generally decreases as the protective corrosion layers form at temperatures above 60 °C [24].

The influence of CO_2 partial pressure. In general, with the increase of CO_2 partial pressure, the pH of the solution decreases, which increases the potential for metal corrosion. Some studies reported that the CO_2 corrosion rate increased with increasing CO_2 partial pressure [19, 43]. However, an increase of CO_2 partial pressure can also thicken the surface film of the steel, which enhances the protection of the steel and reduces the corrosion rate [44]. Therefore, the influence of CO_2 partial pressure on the average corrosion rate is complicated, which needs to be analyzed in combination with other environmental factors like temperature.

The influence of flow rate. Large volumes of CO_2 must be pumped into the deep subsurface for geological CO_2 consumption and storage, and the subsurface flow of CO_2 is a very intricate process. The flow system is a two- or even three-phase system, and it is important to distinguish between the flow in pores and the flow through fractures and wellbores. Most notably, geochemical processes, wettability, and pore structure all play critical roles in controlling subsurface CO_2 flow [45].

The influence of flow rate on CO_2 corrosion is complex. The high flow rate increases the mass transfer rate of the corrosive medium to the metal surface, and the high flow rate hinders the formation of the passivation film on metal surface, so the corrosion rate increases with the increase of the flow rate [46]. According to Dugstad et al. [47] and Srinivasan and Kane [48], in the environment without corrosion inhibitors, when the flow rate is no more than 1 m/s, it can be considered no-corrosion flow. When the flow rate is 1–3 m/s, the film-forming process can still be carried out, and it can be considered limit-corrosion flow. When the flow rate exceeds 4 m/s, the passivation film will be peeled off, revealing a "fresh" layer of the steel, and the corrosion rate is greatly increased. When the flow rate is greater than 5 m/s, the passivation film cannot be formed, and the corrosion rate is the highest. In summary, a high flow rate accelerates CO_2 corrosion of metal materials, which inevitably causes great harm to the tubing and the casing of CCUS wells. As a result, the flow rate should be controlled below 4 m/s.

The influence of pH. In strong acid solutions like HCl solution, the diffusion process of hydrogen ions is a rate-determining step for corrosion reactions, and pH is a key criterion that determines the rate of hydrogen ion diffusion. In short, the corrosivity of strong acid solutions can be directly measured by pH [49]. The corrosion rate decreases with an increasing pH [11, 19, 37, 43]. In CO_2-rich solutions, however, pH should be combined with total dissolved CO_2 as two important factors to evaluate the corrosiveness of the medium. The solubility of CO_2 in water is large, for example, under the condition of CO_2 partial pressure of 0.1 MPa and 25 °C, the solubility of CO_2 in water is 3.5×10^{-2} mol/L. Only a very small portion of the dissolved CO_2 is hydrated to generate H_2CO_3. Therefore, the corrosivity of the aqueous CO_2 solution is not only determined by the pH value of the solution, but also by the concentration of total dissolved CO_2. Researchers have found that under the same pH conditions, the CO_2 aqueous solution is more corrosive than the HCl aqueous solution.

The influence of gas impurities. Through its synergistic interaction with CO_2, H_2S can worsen CO_2 corrosion [50, 51]. Because oxygen can speed up corrosion, O_2 concentrations in enhanced oil recovery (EOR) applications should not go above 1000 ppm. The environment and O_2 concentration have an impact on how oxygen affects corrosion in a supercritical CO_2 environment. O_2 impacts both the rate of corrosion of the steel and the corrosion products when the water content is higher than the critical value. The findings of the experiments indicate that O_2 will significantly raise the local corrosion rate, however it is still debatable how O_2 will affect the overall corrosion rate. Some early studies found that the presence of O_2 accelerated steel casing corrosion due to the catalyzing effect [43, 52, 53]. Some scholars found that O_2 could inhibit the formation of the $FeCO_3$ protective film, which then increased the general corrosion rate. However, some other scholars found that O_2 could passivate the steel and decrease the general corrosion rate [54, 55].

The influence of Cl^- and other dissolved species. The effect of Cl^- is complex, and the effects of Cl^- on passivated steel and non-passivated steel are different. Cl^- itself tends to cause local corrosion such as pit corrosion and crevice corrosion of alloy steel, and hinders the formation of the passivation film. Chen et al. [56] studied the role of Cl^- in CO_2 aqueous solution of API-N80 steel, pointing out that the presence of Cl^- greatly reduces the possibility of passivation film formation. Mao et al. [57] found that the presence of small amount of Cl^- could significantly reduce the passivating tendency of steel, and thus increase the corrosion rate. Qu et al. [58] found that NaCl improved the conductivity of solutions saturated with CO_2, which caused a quick dissolution of protective films on the surface of carbon steel. However, other studies showed that a Cl^- concentration between 1 and 20 wt% had no significant effect on localized corrosion rate at 80 °C [50]. Cl^- not only plays a negative role in steel corrosion. The presence of Cl^- at room temperature can reduce the solubility of CO_2 and may reduce the corrosion rate. It is reported that when $PCO_2 = 5.5$ MPa and the temperature is 150 °C, if the content of NaCl is less than 10%, the corrosion rate of carbon steel decreases slightly with the increase of Cl^- content. However, when the content of NaCl is greater than 10%, the corrosion rate of carbon steel sharply increases with the increase of Cl^- content. Although Cl^- is usually not detected in the corrosion scale, it is regarded to have significant influence on the nucleation and growth of $FeCO_3$ and therefore the properties (e.g., thickness and compactness) of the corrosion scales [24].

A good example of how the dissolved species enhance steel casing corrosion is the corrosion of geothermal wells. Investigation on wellbore casing corrosion is particularly important in maintaining the integrity of geothermal wells, which serve as the key components to ensure the smooth discharge of geothermal fluid from deep subsurface towards the surface. However, most geothermal fluids are corrosive due to the presence of some dissolved species, and the corrosive geothermal fluid can cause severe corrosion on the casing of geothermal wells. The geothermal fluid discharged from geothermal wells contains several dissolved species like hydrochloric acid (HCl), hydrogen sulphide (H_2S), and carbon dioxide (CO_2) [59], which are the primary causes of casing corrosion. Uniform corrosion is the most prevalent corrosion phenomena under any geothermal fluid condition, and is mostly observed on carbon steel casing. Another form of corrosion is H_2S-induced stress cracking, and medium and high-strength carbon and low alloyed steels are susceptible to H_2S-induced stress cracking under various geothermal fluid conditions. Erosion-corrosion occurs in the pipes for two phase (liquid and solid) flow transport which runs at high velocity [60]. To minimize the impacts of aforementioned types of corrosion on geothermal wells, precautions during the design step, and conscious material selection, have an important role. Optimum cost and safety are factors influencing material selection [61]. Currently, the casing of geothermal wells is often designed using carbon steel or low alloyed steel that has high strength but is prone to corrosion. As a result, researchers have tried to use titanium-based alloy, containing the elements nickel, molybdenum, and zirconium to replace carbon steel or low alloyed steel. In previous studies, the titanium-based alloy has been shown to possess a crystal lattice structure with high mechanical stability, and the mechanical stability is well maintained at very high temperatures [59, 60].

The influence of HCO_3^-. The presence of HCO_3^- inhibits the dissolution of $FeCO_3$, promotes the formation of a passivation film on the surface of the steel, and thus reduces the corrosion rate of the carbon steel. The solution with high concentration of HCO_3^- increases the passivation potential interval, increases the breakdown potential, and reduces the chance of pitting corrosion. When HCO_3^- coexists with Ca^{2+}, a protective film is easily formed on the surface of the steel, thereby reducing the corrosion rate.

The influence of hydrocarbons. In the scenario of CO_2 enhanced oil recovery (CO_2-EOR), abundant hydrocarbons are present in the multi-phase system. When the content of hydrocarbons is more than 55 wt%, hydrocarbons inhibit the corrosion of CO_2 on steel, which has little to do with the type of hydrocarbons. When the content of hydrocarbons is less than 55 wt%, the reduction of steel corrosion by hydrocarbons depends on the type of hydrocarbons and the temperature.

The influence of bacteria. Bacterial corrosion is present in all CCUS operations. In CO_2-containing CCUS wells, bacterial corrosion cannot be ignored. Based on the oxygen demand in their growth, bacteria can be divided into aerobic bacteria and anaerobic bacteria. Aerobic bacteria mainly include sulfur-oxidizing bacteria, iron bacteria, and some mucus-forming heterotrophic bacteria. Among them, iron bacteria are the most common. Some of the iron bacteria are autotrophic, growing in a neutral environment. The autotrophic iron bacteria assimilate CO_2 by the energy coming from Fe(II) oxidation. The corrosion effect of aerobic bacteria is mainly divided into two types: one is due to formation of corrosive acids by metabolism, and the other is due to presence of local oxygen batteries. Anaerobes are mainly sulfate-reducing bacteria, and they are the main microorganisms causing metal corrosion hazards, widely distributed in nature, suitable for growing in a temperature range of 30–40 °C, and a pH range of 7–7.5.

The influence of organic acids. Organic acids are often found in CO_2-rich wells at concentrations of a few hundred ppm. The corrosiveness of organic acids depends heavily on the concentration of organic acids and CO_2. The practices of corrosion control of CCUS wells containing organic acids at different CO_2 concentrations demonstrate that the influence of organic acids on corrosion increases with the increase of CO_2 concentration. When a certain level of CO_2 concentration is reached, the influence of organic acids on the corrosion rates becomes not significant.

The influence of inorganic salt scaling. In CCUS operations, inorganic salt scaling may occur in underground reservoirs, wellbores, surface CO_2 storage tanks and transportation pipelines, etc. When CO_2 encounters fluids containing Ca^{2+}, a large amount of $CaCO_3$ may be formed, and $CaCO_3$ scale can be deposited on the surface of the steel pipe, causing local blocking and corrosion under the scale layer.

The influence of external loading. External loading increases the corrosion loss of carbon steel in the CO_2 solution, and the corrosion caused by continuous loading is more serious than that caused by intermittent loading. Zeng et al. [62] have found that the effect of external loads on the time-to-cracking of the steel is noticeable, and the steel corrosion loss varies by ~ 9% under different loading conditions at the surface cracking moment. Furthermore, the higher the load level, the more rapidly the corrosion-induced cracks develop. The maximum corrosion-induced crack width on the surface of the steel bar subjected to 60% ultimate load is 1.22 times of the crack width of the steel bar with no load. In short, the loading and CO_2 have a synergistic effect on the corrosion of steel. The loading and the CO_2 reaction interacting together can produce a severe corrosion effect that is greater than the cumulative effect that the loading and the CO_2 reaction individually produce.

4.4.3 Summary

In summary, CO_2-induced steel corrosion is mainly caused by H^+ from dissociation of H_2CO_3, and the concentration of H_2CO_3 needs to be high to induce noticeable corrosion. Direct reduction of hydrogen in H_2CO_3 and HCO_3^- also contributes to CO_2-induced steel corrosion. Fe^{2+} released from steel surface as a result of corrosion can combine with HCO_3- to produce $FeCO_3$, and the deposition of $FeCO_3$ can form a corrosion product film. The corrosion product film can serve as a protective barrier against corrosion, and the properties of the film are the key to determine the degree of corrosion. Factors like temperature, CO_2 partial pressure, chromium content in steel, presence of H_2S or O_2, Cl^- content, etc. can have a vital effect on the level of CO_2-induced steel corrosion.

4.5 Wellbore Cement Corrosion

Cement and concrete corrosion by CO_2 under atmospheric conditions, usually referred to as carbonation, has been very widely studied by cement and concrete research community [63–65]. In concrete research, carbonation is the term for the reaction between the lime in concrete and the CO_2 in the air, causing formation of $CaCO_3$. Under low CO_2 concentration, CO_2 usually does not cause severe cement and concrete damage. Carbonation under low CO_2 concentration sometimes strengthens cement and concrete, and some researchers intentionally introduce CO_2 to concrete at an early age for a short period of time to promote strength gain of concrete, which is referred to as accelerated carbonation [66]. However, some researchers believe that carbonation reduces concrete quality and its ability to protect steel reinforcement from corrosion, and results in additional shrinkage in the carbonated region [64, 67].

Under very high CO_2 concentration, cement and concrete damage induced by carbonation may become severe. In scenarios with very high CO_2 concentration like geologic CO_2 utilization and storage conditions, the pH becomes very low due to formation of large amounts of carbonic acid, and carbonic acid is able to induce cement corrosion [14, 68]. As a result, wellbore cement materials exposed to high CO_2 concentration under geologic CO_2 utilization and storage conditions are of primary concern. Low pH is able to cause cement corrosion due to the fact that cement is an alkaline material. When an acid comes into contact with cement and concrete, the first line of attack arises from the reaction of the acid with cement hydration products like $Ca(OH)_2$ and C–S–H, causing dissolution of these hydration products. It is generally acknowledged that the dissolution of C–S–H will only follow after portlandite dissolution [69, 70]. The dissolution of C–S–H causes more advanced and severe cement damage, above and beyond the effects arising from portlandite dissolution alone [71]. In short, a typical CO_2-induced cement corrosion process can be divided into three steps:

Step 1: $Ca(OH)_2$ and C–S–H dissolution. Cement hydration products like $Ca(OH)_2$ and calcium silicate hydrate (C–S–H) react with H^+ from carbonic acid and partially get dissolved. The reactions are listed below:

$$Ca(OH)_2 + 2\,H^+ \rightarrow Ca^{2+} + 2\,H_2O \tag{4.20}$$

$$C - S - H + 2\,H^+ \rightarrow Ca^{2+} + SiO_2(s,\ am) + H_2O \tag{4.21}$$

Step 2: $CaCO_3$ precipitation. Dissolution of $Ca(OH)_2$ and C–S–H releases Ca^{2+}, which reacts with dissolved H_2CO_3, HCO_3^- and CO_3^{2-} to form $CaCO_3$ (Reactions 4.22–4.24). With very high dissolved CO_2 concentration, precipitated $CaCO_3$ may form a thick $CaCO_3$ layer.

$$Ca^{2+} + H_2CO_3 \rightarrow CaCO_3(s) + 2H^+ \tag{4.22}$$

$$Ca^{2+} + HCO_3^- \rightarrow CaCO_3(s) + H^+ \tag{4.23}$$

$$Ca^{2+} + CO_3^{2-} \rightarrow CaCO_3(s) \tag{4.24}$$

Step 3: $CaCO_3$ dissolution. $CaCO_3$ formed in Step 2 reacts with H^+ and partially gets dissolved (Reaction 4.16). $CaCO_3$ dissolution typically occurs at the exterior of the $CaCO_3$ layer, where the pH of the aqueous phase is low.

$$CaCO_3 + H^+ \rightarrow Ca^{2+} + HCO_3^- \tag{4.25}$$

Kutchko et al. [14] studied the process of cement corrosion by high concentration CO_2 and they proposed a schematic (Fig. 4.7a) that can describe the three corrosion steps. The corresponding reaction zones can be well matched by micro-CT analysis (Fig. 4.7b).

As shown in Fig. 4.7b, CT imaging of CO_2-induced cement corrosion has clearly demonstrated the reaction patterns of $Ca(OH)_2$ and C–S–H dissolution (zone 1), $CaCO_3$ precipitation (zone 2), and $CaCO_3$ dissolution (zone 3). $Ca(OH)_2$ and C–S–H dissolution, and $CaCO_3$ dissolution cause formation of two high-porosity regions in

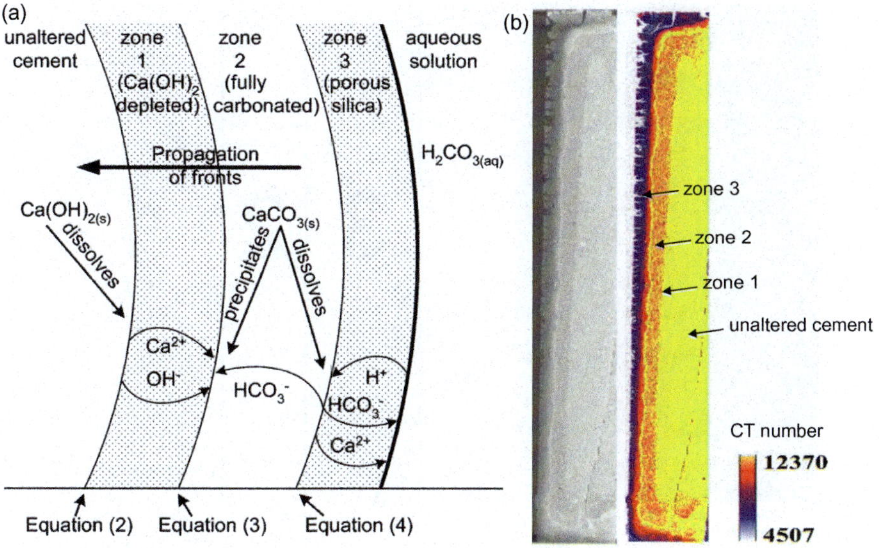

Fig. 4.7 A schematic **a** showing the key reactions between cement and CO_2 [14] and a micro-CT image **b** showing the reaction zones [72]

cement, and $CaCO_3$ precipitation causes formation of a low-porosity region between the two high-porosity regions. This high-porosity → low-porosity → high-porosity region distribution from the surface to the interior of cement looks like a sandwich, and is hereafter referred to as "sandwich pattern". This sandwich pattern has also been observed by SEM. For example, Kutchko et al. [73] exposed Class H neat cement to CO_2-saturated brine for 9 days under the condition of 50 °C and 30.3 MPa CO_2 partial pressure, and the SEM-BSE images of the exposed cement revealed 3 distinct zones, as described above. Sauki and Irawan [74] divided the after-exposure cement sample into four distinct zones (Fig. 4.8). Zone 1 was the innermost unaltered cement. Zone 2 was a 50–100 m-thick zone that had a little higher porosity and lower $Ca(OH)_2$ content. Zone 3 had a 100–200 m-thick ring with decreased porosity and elevated $CaCO_3$ content. The BSE-SEM picture of Fig. 4.6 demonstrated that Zone 3 was less porous than any other locations including the unmodified cement. In all carbonated samples, only calcite and aragonite were discernible from XRD examination, and vaterite was not discovered. Zone 4 (200–400 m in size), which showed the greatest increase in porosity and greatest depletion of calcium as a result of severe cement degradation, was the area that showed the most obvious signs of attack. Zhang et al. [75] prepared pozzolan-amended wellbore cement samples and exposed the samples to mixtures of H_2S and CO_2 under laboratory-simulated geologic CO_2 storage conditions (50 °C and 15.2 MPa) for 2.5, 9, 28 and 90 days, and the distinct 3-zone pattern was clearly observed after 2.5 days of exposure to H_2S and CO_2.

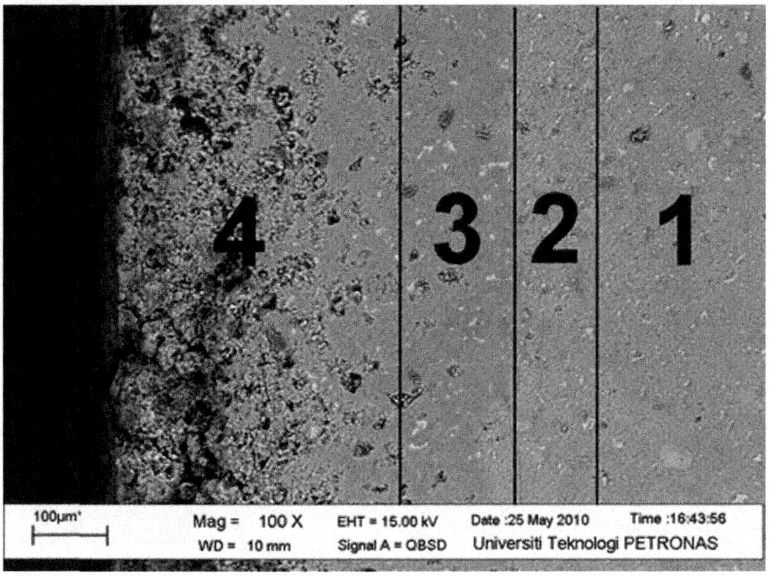

Fig. 4.8 BSE-SEM image of cement degradation after a 120-h-CO_2 attack in brine solution at 140 bar and 40 °C. The locations of 4 distinct zones are numbered [74]

It is generally believed that the formation of a $CaCO_3$ precipitation zone hinders further penetration of CO_2 into the interior of cement, and the $CaCO_3$ precipitation zone is sometimes called a protective layer. For example, Jacquemet et al. [76] conducted an exposure experiment of Class G cement (with silica flour added) at 50 MPa and 120 °C (the samples were immersed in brine saturated with a mixture of H_2S and CO_2) for up to 60 days. They observed armouring of the cement through the fast creation of a dense and non-porous calcite coating, and a global porosity decrease of the cement due to pore clogging. However, some researchers argue that excessive carbonation may cause cement damage [77–81]. For example, Barlet-Gouédard et al. [77] reported a strength loss of up to 65% for heavily carbonated Portland cements after a 42-day exposure to CO_2-saturated brine, and the occurrence of cracks in the carbonation layer during compressive strength measurements. Fabbri et al. [78] observed a significant degradation of mechanical properties of carbonated cement under wet supercritical CO_2 conditions at an elevated temperature. The degradation is obvious with the presence of micro-cracks in the carbonated layers of the Portland cement. Gharardi et al. [80] had a detailed discussion on excessive cement carbonation. They argue that though the deposition of $CaCO_3$ in the pores may locally allow the porosity to decrease, massive carbonation caused by CO_2 corrosion should be considered not only as mechanisms of possible further sealing of the cement, but also as possible sources of late mechanical degradation of the material. In short, whereas moderate carbonation can be beneficial to cement strength, the formation of cracks is usually recognized in association with extensive carbonation processes. Um et al. [81] observed formation of a thick carbonation zone in cement cores after reacting

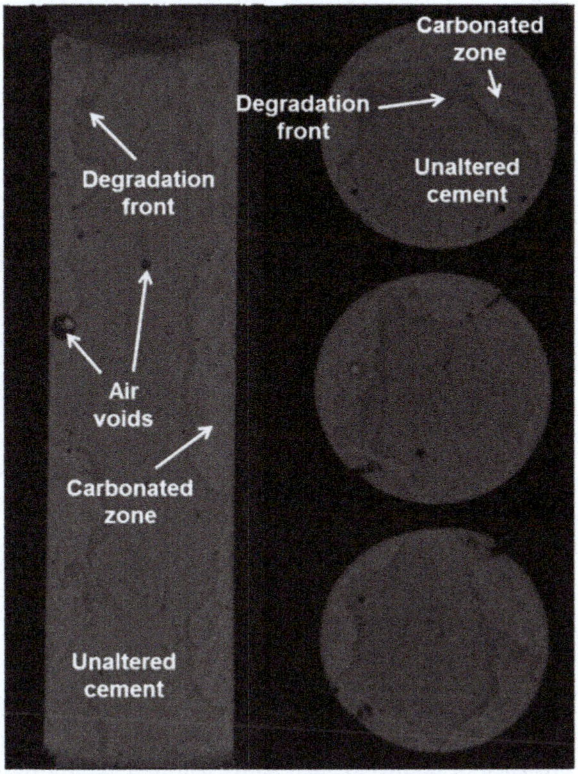

Fig. 4.9 A demonstration of excessive carbonation and loss of cement integrity due to CO_2 attack [81]

with CO_2, and the carbonation zone extended to the interior of cement, causing a visible degradation of cement (Fig. 4.9).

Many factors like water to solid ratio, stress, salinity, external additives, presence of H_2S and SO_4^{2-}, etc. affect the level of cement corrosion. Lu et al. [82] tested the CO_2-resistance of three class G cement samples with different water to solid ratios. They found that the CO_2-induced corrosion depth increased with the increase of water to solid ratio, given the same corrosion time. The level of carbonation in concrete or cement is also affected by external loading. For example, Wang et al. [83] carried out an experimental study to investigate the depth of carbonation in concrete mortar subjected to different levels of non-destructive axial tensile and compressive loads. They found that a higher compressive load made the sample dense and reduce the depth of carbonation, while a higher tensile load made the sample form micro-cracks and cause an increase in the depth of carbonation (Fig. 4.10). As for salinity, the formation water usually has a very high salinity in geological CO_2 storage conditions, and penetration of dissolved Cl^- from formation water into cement becomes possible. As a result, an environment with very high Cl- content favors penetration of Cl^- through cement and the Cl^- may reach steel casing, which can cause casing damage.

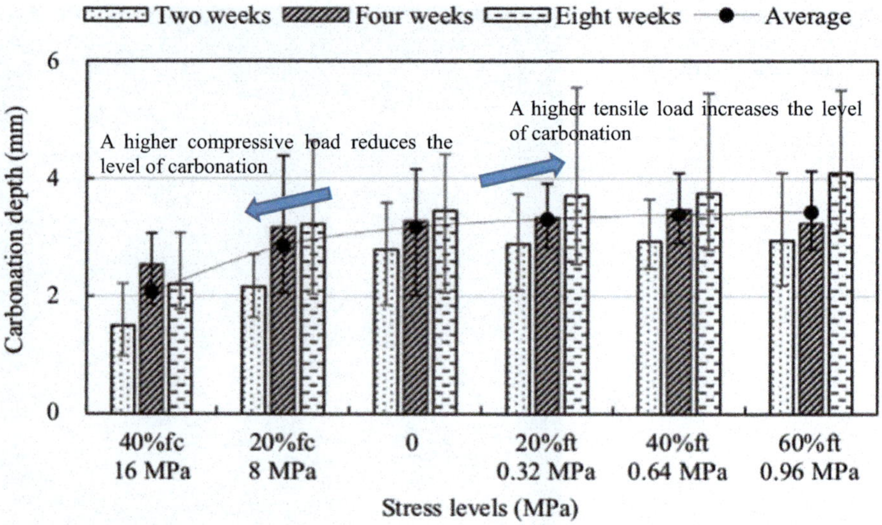

Fig. 4.10 Relationship between carbonation depth and stress for loaded mortar samples. fc = compressive strength of the sample; ft = tensile strength of the sample [83]

The degree of cement corrosion is closely related to the amount of additives in cement [84]. G-class oilwell cement is more resistant to CO_2 corrosion than ordinary Portland cement due to its higher content of $CaSO_4$ and lower iron content. The presence of additives such as fly ash in a certain range in cement can improve the resistance of cement to CO_2 corrosion, but if the amount of additives is too high, the permeability of the cement system may significantly increase, which leads to easy penetration of CO_2 into the interior of cement and a high CO_2 corrosion rate. For example, Kutchko et al. [85] compared the CO_2 corrosion rate of a cement sample with 35 vol% fly ash additive with that of another sample with 65 vol% fly ash additive. The corrosion rate of the 35 vol% fly ash sample was significantly higher than that of the 65 fly ash vol% sample, after 9 days of CO_2 corrosion. This may be a result of the lower slurry density and higher water to cement ratio (which ultimately relates to the permeability of the set cement) for the 65 vol% fly ash sample.

A medium concentration of H_2S can increase the corrosion rate of cement but the degree of corrosion is tolerable [76, 86–88]. For example, Kutchko et al. [88] exposed cement samples to a mixture of H_2S and CO_2 (21 mol% H_2S to simulate acid gas) for a period of 28 days at a temperature of 50 °C and a pressure of 15 MPa. The H_2S–CO_2 exposed cement exhibited a carbonated zone similar to previously reported CO_2-only samples, and the cement also underwent an additional sequence of oxidation–reduction and sulfidation reactions. Ettringite was observed in the interior region of the cement, and pyrite was found in the carbonated rim of the cement. Though the corrosion of cement was observable, the integrity of cement was maintained. However, when the concentration of H_2S is elevated, it can cause very severe cement structural damage. Kutchko et al. [89] exposed a cement sample with 35 vol%

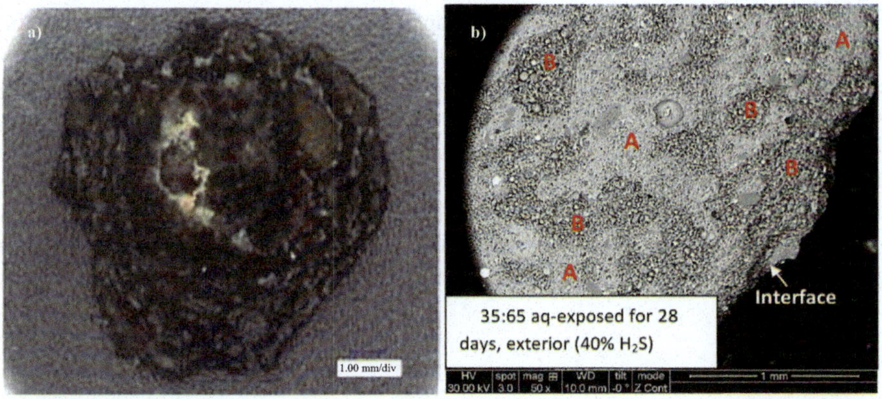

Fig. 4.11 **a** Optical image of a polished thin section of the cement sample with 35 vol% pozzolan additive exposed to pure H_2S (50 °C and 15 MPa) for 28 days [89]. The sample lost much of its structural integrity; **b** SEM image of the cement sample with 35 vol% pozzolan additive exposed to 40 mol% H_2S and 60 mol% CO_2 (50 °C and 15 MPa) for 28 days [90]. A = carbonate-rich zone; B = decalcified zone

pozzolan additive to pure H_2S for a period of 28 days at a temperature of 50 °C and a pressure of 15 MPa. The exposed cement underwent significant physical and chemical alteration via sulfate formation and decalcification, and the sample showed a complete loss of structural integrity (Fig. 4.11a). Zhang et al. [90] exposed a cement sample with 35 vol% pozzolan additive to 40 mol% H_2S and 60 mol% CO_2 for a period of 28 days at a temperature of 50 °C and a pressure of 15 MPa. They found that with the presence of 40 mol% H_2S in the gas mixture, a large region near the exterior of cement was corroded, showing a combination of carbonate-rich zone and decalcified zone (Fig. 4.11b). The degree of structural deterioration for the sample exposed to 40 mol% H_2S was less severe compared with the case of pure H_2S exposure.

The possible interactions between SO_4^{2-} and cement during CO_2 attack under geologic CO_2 storage conditions need to be studied, because SO_4^{2-} concentrations are usually between 0.01 M and 0.05 M and can be as high as 0.15 M even before CO_2 injection in formation waters [91, 92]. Moreover, SO_4^{2-} can be produced by co-injection of H_2S, during which S can be oxidized either by O_2, Fe(III), or other redox species [93]. To investigate the effects of SO_4^{2-} on CO_2 attack on wellbore cement (i.e., chemical and mechanical alterations) during geologic CO_2 storage, Li et al. [92] reacted cement samples in brine with 0.05 M SO_4^{2-} and 0.4 M NaCl at 95 °C under 10 MPa of supercritical CO_2. SEM and inductively coupled plasma optical emission spectrometry (ICP-OES) analyses on reacted samples showed that the CO_2 attack in the presence of additional SO_4^{2-} was much less severe than that in the system without additional SO_4^{2-}. The results from three-point bending tests also indicated that SO_4^{2-} substantially mitigated the deterioration of the strength and elastic modulus of reacted cement samples. Furthermore, typical SO_4^{2-} attacks on cement via formation of ettringite were not observed [92].

In summary, cement and concrete damage induced by CO_2 corrosion may become severe due to very low pH in geologic CO_2 utilization and storage environment. The surface of wellbore cement in CO_2 utilization and storage environment generally experiences dissolution. However, the expansion of the dissolution zone is hindered by formation of a dense calcium carbonate layer. In most cases, the calcium carbonate layer is protective, but excessive carbonation may cause cement damage. Many factors like water to solid ratio, stress, salinity, external additives, presence of H_2S, etc. affect the level of cement corrosion, and the presence of H_2S should be the primary concern, because the corrosion rate of wellbore cement is quite sensitive to H_2S content.

4.6 Corrosion of Monitoring Devices

A thorough and carefully planned monitoring strategy and trustworthy monitoring findings are required to assure the safety of storing CO_2 in deep subterranean formations, which are crucial in convincing the general public to embrace CCUS technology [94]. Many monitoring technologies have been used in large-scale CCUS projects worldwide, including seismic monitoring, CO_2 flux monitoring, CO_2 concentration monitoring, pressure monitoring, pH monitoring, heavy metal monitoring, differential interferometric synthetic aperture radar (D-InSAR) monitoring, etc. [94].

The monitoring devices that are in direct contact with high concentration CO_2 are prone to corrosion. One example is pressure sensors. Pressure sensors are usually placed in monitoring wells to record the pressure changes at a given subsurface formation. The pressure change is an important criterion to evaluate the injectivity of CO_2 into the target CO_2 storage formation, and the pressure change can also be used to evaluate the potential for the CO_2 to breakthrough the caprock. Pressure sensors need to be in direct contact with CO_2-rich fluid to record the pressure changes. The CO_2 and other corrosive substances in the fluid may damage the sensor-making materials, including metals and organic compounds to make sealing rings. As a result, unique challenges are raised for the design of pressure sensors with CO_2-rich fluid compatibility. To overcome corrosion, corrosion-resistant materials need to be used to make pressure sensors for CCUS applications. For use with CO_2-rich fluid, corrosion-resistant pressure sensors frequently use housings composed of stainless steel or plastics like PVDF, PVC, or PPS. Most sensors employ ceramic sensor elements, but some also use silicon oil-backed stainless steel diaphragms to transmit pressure to the sensing element [95]. The O-ring that connects the pressure sensor's body to the ceramic diaphragm can be manufactured of a variety of materials, including fluorocarbon plastic, nitrile rubber, ethylene propylene diene monomer, etc. In some countries, corrosion-resistant sensors require certification by standards like IECEx 02 (for the entire world) and ATEX 95 (for Europe). Under EU law, ATEX 95 is required for all electrical and non-electrical equipment that is used in hazardous environments, while IECEx 02 is intended only for electrical equipment in hazardous environments [95].

Another example is CO_2 flux monitoring device. A shallow subsurface CO_2 flux monitoring is a direct indicator of whether the injected CO_2 leaks to the surrounding environment. At Shenhua CCS site, the CO_2 flux at a height of 10 m below the ground is measured using an in-situ underground CO_2 flux analyzer, from which the CO_2 flux in the real situation is obtained as it can suppress any errors caused by ambient gas transports in the conventional measurements [94]. The CO_2 concentration at shallow subsurface is much lower than the CO_2 concentration at the CO_2 injection formation, so the corrosivity of the shallow subsurface is less severe than that at the CO_2 injection formation. As a result, for the CO_2 flux monitoring device, ordinary corrosion control measures can be applied, such as painting of coatings on the surface of metal components in direct contact with the surrounding environment. There is no need to use corrosion-resistant materials to make the bulk body and the sealing rings of the in-situ underground CO_2 flux analyzer.

References

1. American Petroleum Institute: API RP 90: Annular casing pressure management for offshore wells, Washington: American Petroleum Institute, 1–96 (2006)
2. Norwegian Oil and Gas Association and The Federation of Norwegian Industries: NORSOK Standard D-010: Well integrity in drilling and well operations, Norway, NORSOK, 1–224 (2013)
3. Saint-Marc, J., Garnier, A., Bois, A.P.: Initial state of stress: the key to achieving long-term cement-sheath integrity, SPE Annual Technical Conference and Exhibition, Denver, U.S., SPE-116651-MS, OnePetro (2008)
4. Kalousek, G.L.: The reactions of cement hydration at elevated temperatures. In: Proceedings of the Third International Symposium on the Chemistry of Cement, London, UK, pp. 334–355, ICE Publishing (1952)
5. Nelson, E., Eilers, L.: Cementing steamflood and fireflood wells-slurry design. J. Can. Pet. Technol. **24**(05), PETSOC-85-05-06 (1985)
6. Wang, Y.B., Gao, D.L., Fang, J.: Assessment on the cement mechanical properties of shale gas horizontal well in fracturing process. Petroleum Tubular Goods & Instruments **6**(4), 5 (2020)
7. Gao, D.L., Wang, Y.B., Fang, J.: Mechanical characteristics analysis of casing-cement ring-formation multilayer composite system. CJAM **31**(3), 6 (2014)
8. Li, J., Guo, X.L., Liu, G.H.: The influence of cement sheath property on casing stress considering thermo-pressure coupling effect. Science Technology and Engineering **18**(13), 8 (2018)
9. Li, T., Guan, Z.C., Li, J.C., Zhao, X.F., Xia, Y.: Matching relationship between casing-cement sheath-stratum in gas storages. Transportation & Storage **38**(11), 6 (2019)
10. Zinkham, R.E., Goodwin, R.J.: Burst resistance of pipe cemented into the earth. T. Soc. Petrol. En. Aime. **14**(09), 1033–1040 (1962)
11. Waard, C.D., Milliams, D.: Carbonic acid corrosion of steel. Corrosion **31**(5), 177–181 (1975)
12. Ogundele, G., White, W.: Observations on the influences of dissolved hydrocarbon gases and variable water chemistries on corrosion of an API-L80 steel. Corrosion **43**(11), 665–673 (1987)
13. Sim, S., Cavanaugh, M.K., Corrigan, P., Cole, I., Birbilis, N.: Aqueous corrosion testing and neural network modeling to simulate corrosion of supercritical CO_2 pipelines in the carbon capture and storage cycle. Corrosion **69**(5), 477–486 (2013)

14. Kutchko, B.G., Strazisar, B.R., Dzombak, D.A., Lowry, G.V., Thaulow, N.: Degradation of well cement by CO_2 under geologic sequestration conditions. Environ. Sci. Technol. **41**(13), 4787–4792 (2007)
15. Xu, L., Xu, J., Xu, M.B., Li, S.Y., Liu, S., Huang, Y., You, F.C.: Corrosion behavior of 3% Cr casing steel in CO-containing environment: a case study. The Open Petroleum Engineering Journal **11**(1), 1–13 (2018)
16. Remita, E., Tribollet, B., Sutter, E., Vivier, V., Ropital, F., Kittel, J.: Hydrogen evolution in aqueous solutions containing dissolved CO_2: quantitative contribution of the buffering effect. Corros. Sci. **50**(5), 1433–1440 (2008)
17. Kahyarian, A., Brown, B., Nešić, S.: The unified mechanism of corrosion in aqueous weak acids solutions: a review of the recent developments in mechanistic understandings of mild steel corrosion in the presence of carboxylic acids, carbon dioxide, and hydrogen sulfide. Corrosion **76**(3), 268–278 (2020)
18. Kahyarian, A.: Mechanism and Prediction of Mild Steel Corrosion in Aqueous Solutions Containing Carboxylic Acids, Carbon Dioxide, and Hydrogen Sulfide. Ohio University, Russian College of Engineering and Technology, Columbus (2018)
19. Nesic, S., Postlethwaite, J., Olsen, S.: An electrochemical model for prediction of corrosion of mild steel in aqueous carbon dioxide solutions. Corrosion **52**(4), 280–294 (1996)
20. Gulbrandsen, E., Bilkova, K.: Solution chemistry effects on corrosion of carbon steels in presence of CO_2 and acetic acid. In: Corrosion 2006: National Association of Corrosion Engineers International Annual Conference and Corrosion Show, San Diego, U.S., NACE-06364. OnePetro (2006)
21. Mansoori, H., Brown, B., Young, D., Nešić, S., Singer, M.: Effect of $Fe_xCa_yCO_3$ and $CaCO_3$ scales on the CO_2 corrosion of mild steel. Corrosion **75**(12), 1434–1449 (2019)
22. Rezaee, A., Nejad, A.M., Mansouri, H., Sciences, A.: Scaling and corrosion in oil production-how do they relate to each other? IJEAS **4**(6), 257443 (2017)
23. Hou, B., Zhang, Q., Li, Y., Zhu, G., Zhang, G.: Influence of corrosion products on the inhibition effect of pyrimidine derivative for the corrosion of carbon steel under supercritical CO_2 conditions. Corros. Sci. **166**, 108442 (2020)
24. Li, S., Zeng, Z., Harris, M.A., Sánchez, L.J., Cong, H.J.: CO_2 corrosion of low carbon steel under the joint effects of time-temperature-salt concentration. Front. Mater. **6**, 10 (2019)
25. Chen, C., Lu, M., Sun, D., Zhang, Z., Chang, W.: Effect of chromium on the pitting resistance of oil tube steel in a carbon dioxide corrosion system. Corrosion **61**(06), NACE-05060594 (2005)
26. Dong, B., Liu, W., Zhang, Y., Banthukul, W., Zhao, Y., Zhang, T., Fan, Y., Li, X.: Comparison of the characteristics of corrosion scales covering 3Cr steel and X60 steel in CO_2-H_2S coexistence environment. J. Nat. Gas. Sci. Eng. **80**, 103371 (2020)
27. Yin, Z., Zhang, Y., Chang, G., Yang, T.: Corrosion behavior and characteristics of 3Cr steel in coexisting H_2S-and CO_2-containing solutions. J. Mater. Eng. Perform. **29**(8), 5442–5457 (2020)
28. Koguma, K., Ikeda, M., Kaku, H., Matsuda, K., Kaneta, M.: Localized corrosion of MoS2 sputtered films in atmosphere-measurement of corrosion current in deionized water. J. Jpn. Soc. Tribol. **50**(7), 546–552 (2005)
29. Sun, B.Z., Zuo, X.M., Cheng, X.Q., Li, X.G.: The role of chromium content in the long-term atmospheric corrosion process. NPJ Mater. Degrad. **4**(1), 1–9 (2020)
30. Hua, Y., Barker, R., Neville, A.: The influence of SO_2 on the tolerable water content to avoid pipeline corrosion during the transportation of supercritical CO_2. Int. J. Greenh. Gas. Con. **37**, 412–423 (2015)
31. Cui, G., Yang, Z.Q., Liu, J.G., Li, Z.L.: A comprehensive review of metal corrosion in a supercritical CO_2 environment. Int. J. Greenh. Gas. Con. **90**, 102814 (2019)
32. Ayello, F., Evans, K., Sridhar, N., Thodla, R.: Effect of liquid impurities on corrosion of carbon steel in supercritical CO_2. In: 8th International Pipeline Conference, Calgary, Canada, pp. 111–123. The American Society of Mechanical Engineers (2010)

33. Choi, Y.S., Nešić, S.: Determining the corrosive potential of CO_2 transport pipeline in high pCO_2–water environments. Int. J. Greenh. Gas. Con. **5**(4), 788–797 (2011)
34. Choi, Y.S., Farelas, F., Nešić, S., Magalhães, A.A.O., Azevedo, A.D.C.: Corrosion behavior of deep water oil production tubing material under supercritical CO_2 environment: part 1—effect of pressure and temperature. Corrosion **70**(1), 38–47 (2014)
35. Rosli, N.R., Choi, Y.S., Young, D.: Impact of oxygen ingress in CO_2 corrosion of mild steel. In: Corrosion 2014: National Association of Corrosion Engineers International Annual Conference and Corrosion Show, San Antonio, U.S., NACE-2014-4299. OnePetro (2014)
36. Ezuber, H.M.: Influence of temperature and thiosulfate on the corrosion behavior of steel in chloride solutions saturated in CO_2. Mater. Design **30**(9), 3420–3427 (2009)
37. Videm, K., Dugstad, A.: Corrosion of carbon steel in an aqueous carbon dioxide environment. Mater. Performance **28**(4), 6238710 (1989)
38. Mishra, B., Al-Hassan, S., Olson, D., Salama, M.: Development of a predictive model for activation-controlled corrosion of steel in solutions containing carbon dioxide. Corrosion **53**(11), 852–859 (1997)
39. Kermani, M., Morshed, A.: Carbon dioxide corrosion in oil and gas productionA compendium. Corrosion **59**(08), NACE-03080659 (2003)
40. Tang, L., Mu, G., Liu, G.: The effect of neutral red on the corrosion inhibition of cold rolled steel in 1.0 M hydrochloric acid. Corros. Sci. **45**(10), 2251–2262 (2003)
41. Shukla, S.K., Quraishi, M.: 4-Substituted anilinomethylpropionate: new and efficient corrosion inhibitors for mild steel in hydrochloric acid solution. Corros. Sci. **51**(9), 1990–1997 (2009)
42. Porcayo, C.J., de la Escalera, L.M., Canto, J., Casales-Diaz, M., Salinas, B.V.J.: Effect of the temperature on the CO_2-corrosion of Ni_3Al. Int. J. Electrochem. Sci. **10**(3), 3136–3151 (2015)
43. De Waard, C., Lotz, U., Milliams, D.: Predictive model for CO_2 corrosion engineering in wet natural gas pipelines. Corrosion **47**(12), 976–985 (1991)
44. Kinsella, B., Tan, M.Y., Bailey, S.: Studies of CO_2 corrosion product scales using electrochemical impedance spectroscopy and surface characterisation techniques. Corrosion: J. Sci. Eng. **54**(10), 8 (1998)
45. Zhang, L., Chen, L., Hu, R., Cai, J.: Subsurface multiphase reactive flow in geologic CO_2 storage: key impact factors and characterization approaches. AGER **6**(3), 179–180 (2022)
46. Lunde, L.: Carbon dioxide corrosion of carbon steel in two-phase flow. Corrosion **50**(09), NACE-94090717 (1994)
47. Dugstad, A., Lunde, L., Videm K.: Parametric study of CO_2 corrosion of carbon steel. In: Corrosion 94: National Association of Corrosion Engineers International Annual Conference and Corrosion Show, Houston, U.S., 5005p. OnePetro (1994)
48. Srinivasan, S., Kane, R.D.: Prediction of corrosivity of CO_2/H_2S production environments. In: Corrosion 96: National Association of Corrosion Engineers International Annual Conference and Corrosion Show, Denver, U.S., NACE-96011. OnePetro (1996)
49. Kiani, S., Jafari, S., Apourvari, S.N., Mehrjoo, H.J.: Simulation study of wormhole formation and propagation during matrix acidizing of carbonate reservoirs using a novel in-situ generated hydrochloric acid. AGER **5**(1), 64–74 (2021)
50. Liu, Q., Mao, L., Zhou, S.: Effects of chloride content on CO_2 corrosion of carbon steel in simulated oil and gas well environments. Corros. Sci. **84**, 165–171 (2014)
51. Choi, Y.S., Hassani, S., Vu, T.N., Nešić, S., Abas, A.Z.B.: Effect of H_2S on the corrosion behavior of pipeline steels in supercritical and liquid CO_2 environments. Corrosion **72**(8), 999–1009 (2016)
52. Khokhar, M., Allam, I., Quddus, A.: The role of corrosion in the deposition of $SrCO_3$ crystals. Corrosion **47**(5), 341–343 (1991)
53. Videm, K., Koren, A.: Corrosion, passivity, and pitting of carbon steel in aqueous solutions of HCO_3^-, CO_2, and Cl^-. Corrosion **49**(9), 746–754 (1993)
54. Sun, C., Sun, J., Wang, Y., Lin, X., Li, X., Cheng, X., Liu, H.: Synergistic effect of O_2, H_2S and SO_2 impurities on the corrosion behavior of X65 steel in water-saturated supercritical CO_2 system. Corros. Sci. **107**, 193–203 (2016)

55. Tang, Y., Guo, X., Zhang, G.: Corrosion behaviour of X65 carbon steel in supercritical-CO_2 containing H_2O and O_2 in carbon capture and storage (CCS) technology. Corros. Sci. **118**, 118–128 (2017)
56. Chen, C.F., Lu, M.X., Zhao, G.X., Bai, Z.Q., Yan, M.L., Yang, Y.Q.: Effects of temperature, Cl-concentration and Cr on electrode reactions of CO_2 corrosion of N80 steel. Acta. Metall. Sin. **39**(8), 848–854 (2003)
57. Mao, X., Liu, X., Revie, R.: Pitting corrosion of pipeline steel in dilute bicarbonate solution with chloride ions. Corrosion **50**(09), NACE-94090651 (1994)
58. Qu, Q., Ma, J., Wang, L., Li, L., Bai, W., Ding, Z.: Corrosion behaviour of AZ31B magnesium alloy in NaCl solutions saturated with CO_2. Corros. Sci. **53**(4), 1186–1193 (2011)
59. Thorhallsson, A.I., Karlsdottir, S.N.: Corrosion behaviour of titanium alloy and carbon steel in a high-temperature, single and mixed-phase, simulated geothermal environment containing H_2S, CO_2 and HCl. Corros. Mater. Degrad. **2**(2), 190–209 (2021)
60. Nogara, J., Zarrouk, S.J.: Corrosion in geothermal environment: part 1: fluids and their impact. Renew. Sust. Energ. Rev. **82**, 1333–1346 (2018)
61. Kaya, T., Hoshan, P.: Corrosion and material selection for geothermal systems. In: Proceedings World Geothermal Congress 2005, Antalya, Turkey, pp. 24–29. Springer Nature (2005)
62. Zeng, B., Wang, Y., Gong, F., Maekawa, K.: Corrosion-induced cracking pattern analysis of RC beam under sustained load considering the poromechanical characteristics of corrosion products. Buildings **12**(12), 2256 (2022)
63. Roy, S., Poh, K., Northwood, D.: Environment: durability of concrete—accelerated carbonation and weathering studies. Build. Environ. **34**(5), 597–606 (1999)
64. Somayaji, S.: Civil Engineering Materials. Butterworth-Heinemann, Oxford (2006)
65. Yoon, I.S., Chang, C.H.: Time evolution of CO_2 diffusivity of carbonated concrete. Appl. Sci. Basel. **10**(24), 8910 (2020)
66. Shao, Y., Rostami, V., He, Z., Boyd, A.J: Accelerated carbonation of Portland limestone cement. J. Mater. Civ. Eng. **26**(1), 117–124 (2014)
67. Chen, Y., Liu, P., Yu, Z.: Effects of environmental factors on concrete carbonation depth and compressive strength. Materials **11**(11), 2167 (2018)
68. Miao, X., Zhang, L., Wang, Y., Wang, L., Fu, X., Gan, M., Li, X.: Composites: characterisation of wellbore cement microstructure alteration under geologic carbon storage using X-ray computed micro-tomography: a framework for fast CT image registration and carbonate shell morphology quantification. Cem. Concr. Compos. **108**, 103524 (2020)
69. Sersale, R., Frigione, G., Bonavita, L.: Acid depositions and concrete attack: main influences. Cem. Concr. Res. **28**(1), 19–24 (1998)
70. Hill, J., Byars, E., Sharp, J., Lynsdale, C., Cripps, J., Zhou, Q.: An experimental study of combined acid and sulfate attack of concrete. Cem. Concr. Compos. **25**(8), 997–1003 (2003)
71. Ekolu, S., Diop, S., Azene, F., Mkhize, N.: Disintegration of concrete construction induced by acid mine drainage attack. J. S. Afr. Inst. Civ. Eng. **58**(1), 34–42 (2016)
72. Gan, M., Zhang, L., Wang, Y., Mei, K., Fu, X., Cheng, X., Bai, M., Liu, H., Li, X.: 3D microstructural changes of an artificial flow channel in wellbore cement under geologic CO_2 storage conditions: Combined effect of effective stress and flow. Constr. Build. Mater. **325**, 126761 (2022)
73. Kutchko, B.G., Strazisar, B.R., Lowry, G.V., Dzombak, D.A., Thaulow, N.: Rate of CO_2 attack on hydrated class H well cement under geologic sequestration conditions. Environ. Sci. Technol. **42**(16), 6237–6242 (2008)
74. Sauki, A., Irawan, S.: Effects of pressure and temperature on well cement degradation by supercritical CO_2. IJET **1**(04), 53–61 (2010)
75. Zhang, L., Dzombak, D.A., Nakles, D.V., Hawthorne, S.B., Miller, D.J., Kutchko, B.G., Lopano, C.L., Strazisar, B.R.: Rate of H_2S and CO_2 attack on pozzolan-amended Class H well cement under geologic sequestration conditions. Int. J. Greenh. Gas. Con. **27**, 299–308 (2014)
76. Jacquemet, N., Pironon, J., Lagneau, V., Saint-Marc, J.: Armouring of well cement in H_2S–CO_2 saturated brine by calcite coating—experiments and numerical modelling. Appl. Geochemistry **27**(3), 782–795 (2012)

77. Barlet, G.V., Rimmelé, G., Goffé, B., Porcherie, O.: Well technologies for CO_2 geological storage: CO_2-resistant cement. Oil Gas Sci. Technol. **62**(3), 325–334 (2007)
78. Fabbri, A., Corvisier, J., Schubnel, A., Brunet, F., Goffé, B., Rimmele, Barlet, G.V.: Effect of carbonation on the hydro-mechanical properties of Portland cements. Cem. Concr. Res. **39**(12), 1156–1163 (2009)
79. Wigand, M., Kaszuba, J.P., Carey, J.W., Hollis, W.K.: Geochemical effects of CO_2 sequestration on fractured wellbore cement at the cement/caprock interface. Chem. Geol. **265**(1–2), 122–133 (2009)
80. Gherardi, F., Audigane, P., Gaucher, E.C.: Predicting long-term geochemical alteration of wellbore cement in a generic geological CO_2 confinement site: tackling a difficult reactive transport modeling challenge. J. Hydrol. **420**, 340–359 (2012)
81. Um, W., Rod, K.A., Jung, H.B., Brown, C.F.: Geochemical alteration of wellbore cement by CO_2 or CO_2+H_2S reaction during long-term carbon storage. Greenh. Gases **7**(5), 852–865 (2017)
82. Lu, P.Q., Liu, R.G., Yang, G.G.: Methods of strengthening anti-CO_2 corrosion of oil well cement stone. J. Mater. Sci. Eng. **38**(4), 566–570 (2020)
83. Wang, Y., Jiang, X., Wang, S., Yang, W., Liu, W., Xing, F., Yang, K., Basheer, P.: Influence of axial loads on CO_2 and Cl^- transport in concrete phases: paste, mortar and ITZ. Constr. Build. Mater. **204**, 875–883 (2019)
84. Laudet, J.B., Garnier, A., Neuville, N., Le, G.Y., Fourmaintraux, D., Rafai, N., Burlion, N., Shao, J.F.: The behavior of oil well cement at downhole CO_2 storage conditions: static and dynamic laboratory experiments. Energy Procedia **4**, 5251–5258 (2011)
85. Kutchko, B.G., Strazisar, B.R., Huerta, N., Lowry, G.V., Dzombak, D.A., Thaulow, N.: CO_2 reaction with hydrated class H well cement under geologic sequestration conditions: effects of flyash admixtures. Environ. Sci. Technol. **43**(10), 3947–3952 (2009)
86. Bachu, S., Gunter, W.D.: Acid-gas injection in the Alberta basin, Canada: a CO_2-storage experience. Geol. Soc. Spec. Publ. **233**(1), 225–234 (2004)
87. Bachu, S., Watson, T.L.: Review of failures for wells used for CO_2 and acid gas injection in Alberta. Canada, Energy Procedia **1**(1), 3531–3537 (2009)
88. Kutchko, B.G., Strazisar, B.R., Hawthorne, S.B., Lopano, C.L., Miller, D.J., Hakala, J.A., Guthrie, G.D.: H_2S–CO_2 reaction with hydrated Class H well cement: acid-gas injection and CO_2 Co-sequestration. Int. J. Greenh. Gas. Con. **5**(4), 880–888 (2011)
89. Kutchko, B.G., Strazisar, B.R., Hawthorne, S.B., Lopano, C.L., Miller, D.J., Hakala, J.A., Guthrie, G.D.: Impact of oil well cement exposed to H_2S saturated fluid and gas at high temperatures and pressures: implications for acid gas injection and co-sequestration. Int. J. Greenh. Gas. Con. **3**(1), 80–101 (2015)
90. Zhang, L.W., Dzombak, D.A., Nakles, D.V., Hawthorne, S.B., Miller, D.J., Kutchko, B., Lopano, C., Strazisar, B.: Effect of exposure environment on the interactions between acid gas (H_2S and CO_2) and pozzolan-amended wellbore cement under acid gas co-sequestration conditions. Int. J. Greenh. Gas. Con. **27**, 309–318 (2014)
91. Keller, S.J.: Analyses of Subsurface Brines of Indiana. Indiana Geological & Water Survey, Bloomington (1983)
92. Li, Q., Lim, Y.M., Jun, Y.S.: Effects of sulfate during CO_2 attack on Portland cement and their impacts on mechanical properties under geologic CO_2 sequestration conditions. Environ. Sci. Technol. **49**(11), 7032–7041 (2015)
93. Zhang, L., Dzombak, D.A., Nakles, D.V., Hawthorne, S.B., Miller, D.J., Kutchko, B.G., Lopano, C.L., Strazisar, B.R.: Characterization of pozzolan-amended wellbore cement exposed to CO_2 and H_2S gas mixtures under geologic carbon storage conditions. Int. J. Greenh. Gas. Con. **19**, 358–368 (2013)
94. Zhao, X.L., Ma, R., Zhang, F., Zhong, Z.C., Wang, B.D., Wang, Y.S., Li, Y.L., Weng, L.: The latest monitoring progress for Shenhua CO_2 storage project in China. Int. J. Greenh. Gas. Con. **60**, 199–206 (2017)
95. Avnet Abacus: Pressure sensors for corrosive liquids and gases. https://www.avnet.com/wps/portal/abacus/solutions/technologies/sensors/pressure-sensors/media-types/corrosive-liquids-and-gases/

Chapter 5
Corrosion Control (I): Corrosion-Resistant Steel and Cement

Liwei Zhang, Kaiyuan Mei, Xiaowei Cheng, and Yongcun Feng

5.1 Corrosion-Resistant Steel

As discussed in previous chapters, normal steel such as N80 steel is prone to CO_2-induced corrosion, especially in a humid environment. To mitigate CO_2-induced corrosion, a practical approach is to replace normal steel with CO_2-resistant steel alloys. The most common CO_2-resistant steel alloy is chromium (Cr)-bearing steel alloy, which has a Cr content of 2–27 wt%. In general, the higher the Cr content, the stronger the resistance of the alloy to CO_2 attack. For example, Guo [1] tested the CO_2-resisting performance of 1, 2 and 3 wt% Cr-bearing steel alloys. The commercial API X65 pipeline steel was used for comparison. All the samples were immersed in CO_2-saturated salty water. All tests were performed at 80 °C under stagnant condition, and the CO_2 partial pressure in equilibrium with the CO_2-saturated salty water was 0.8 MPa. They found that the average corrosion rates of X65, 1 wt% Cr, 2 wt% Cr and 3 wt% Cr steels after 517 h of immersion were 9.83 mm/year, 2.40 mm/year, 1.50 mm/year and 1.37 mm/year, respectively. Wu [2] studied corrosion behavior of two types low-alloy steel containing 1 wt% (ferritic-pearlitic microstructure and tempered martensitic microstructure) in CO_2 environments at 60 °C. N80 carbon steel was used for comparison. For N80 carbon steel, severe localized corrosion was observed, and the corrosion rate reached 8.36 mm/year. However, the severe corrosion

L. Zhang
State Key Laboratory of Geomechanics and Geotechnical Engineering, Institute of Rock and Soil Mechanics, Chinese Academy of Sciences, Wuhan 430071, China

University of Chinese Academy of Sciences, Beijing 100049, China

K. Mei (✉) · X. Cheng
School of New Energy and Materials, Southwest Petroleum University, Chengdu 610500, China
e-mail: mky0101@126.com

Y. Feng
College of Petroleum Engineering, China University of Petroleum (Beijing), Beijing 102249, China

L. Zhang (ed.), *Corrosion in CO_2 Capture, Transportation, Geological Utilization and Storage*, Engineering Materials, https://doi.org/10.1007/978-981-99-2392-2_5

did not exist in 1 wt% Cr steel, due to the formation of a compact and self-repairable Cr-rich scale. Also, the corrosion resistance of 1 wt% Cr steel with ferritic-pearlitic microstructure was better than that with tempered martensitic microstructure. The corrosion rate of 1 wt% Cr steel with ferritic-pearlitic microstructure was determined to be 1.99 mm/year, lower than the corrosion rate (3.56 mm/year) of 1 wt% Cr steel with tempered martensitic microstructure. The reason is that the corrosion scale had a worse adherence to the surface of 1 wt% Cr steel with tempered martensitic microstructure, compared with 1 wt% Cr steel with ferritic-pearlitic microstructure (Fig. 5.1). Sun [3] tested the corrosion-resisting performances of tube steels containing 0.25, 2.99 and 5 wt% chromium in 3.5 wt% NaCl solution containing CO_2 and H_2S ($PCO_2 = 5$ MPa, $PH_2S = 0.2$ MPa, $T = 90$ °C, and corrosion time = 360 h). They found that the average corrosion rate of 2.99 wt% Cr was the lowest among the three samples, while the average corrosion rate of the 5 wt% Cr case was the highest. They reach a conclusion that for the Cr-containing steel, the formation of uniform corrosion scales inhibits localized corrosion, but Cr does not substantially improve the resistance to uniform corrosion. In summary, the corrosion rate decreased with increasing Cr content in steels in most studies, with very few exceptions. However, since the cost of Cr-bearing alloy increases with the increase of Cr content, a balance needs to be reached when picking the appropriate Cr content for CO_2-resisting applications.

The existence of Cr in alloys accelerates formation of a corrosion-resistant layer. For alloys containing 3 wt% Cr, Kermani [4] discovered that the corrosion protection is provided by the development of Cr-rich $FeCO_3$ corrosion products (Cr (oxy)hydroxide). According to Sun [5], the primary components of the corrosion scale on X65 steel with 3 wt% Cr were amorphous $Cr(OH)_3$ and $FeCO_3$. Guo [1] reached a similar conclusion that the growth of a passivation layer rich in $Cr(OH)_3$ could effectively hinder the CO_2 induced corrosion on steel (Fig. 5.2). Figure 5.2 shows that after 222 h of reaction, the growth of a passivation layer rich in $Cr(OH)_3$ almost covered the entire surface of the alloy, and CO_2-induced corrosion was significantly inhibited. Though researchers have not reached a widely-accepted conclusion regarding the form of the amorphous Cr-bearing scale, the production rate of the amorphous Cr-bearing scale definitely plays a key role in governing the rate of general corrosion and susceptibility to localized corrosion for Cr-bearing alloys. Besides Cr-bearing scale, possible formation of other scales like crystal $Ca_xFe_{1-x}CO_3$ and amorphous $Ca_xFe_{1-x}CO_3$ further hinders CO_2-induced corrosion (Fig. 5.3).

To further enhance the resistance of Cr-bearing alloys to the attack by CO_2 and other sour gases, other metals like nickel (Ni), molybdenum (Mo) and titanium (Ti) have been added to the alloys. For example, S13Cr alloys, referred to as Super 13Cr, Hyper 13Cr or Modified 13Cr, are alloys with Ni contents in the range of 1–6 wt%, and with molybdenum in the range of 1–2 wt%. The resistance of S13Cr alloys to the attack by CO_2 and other sour gases is generally stronger than the basic API 13Cr grade, and the S13Cr alloys are suitable up to about 30 °C higher operating temperature than the standard 13Cr grades [7]. Also, high temperature halide resistance increases by improving the magnitudes of nickel and chromium in the alloy. Titanium offers ideal resistance to sulphur dioxide, hydrogen sulphide, and

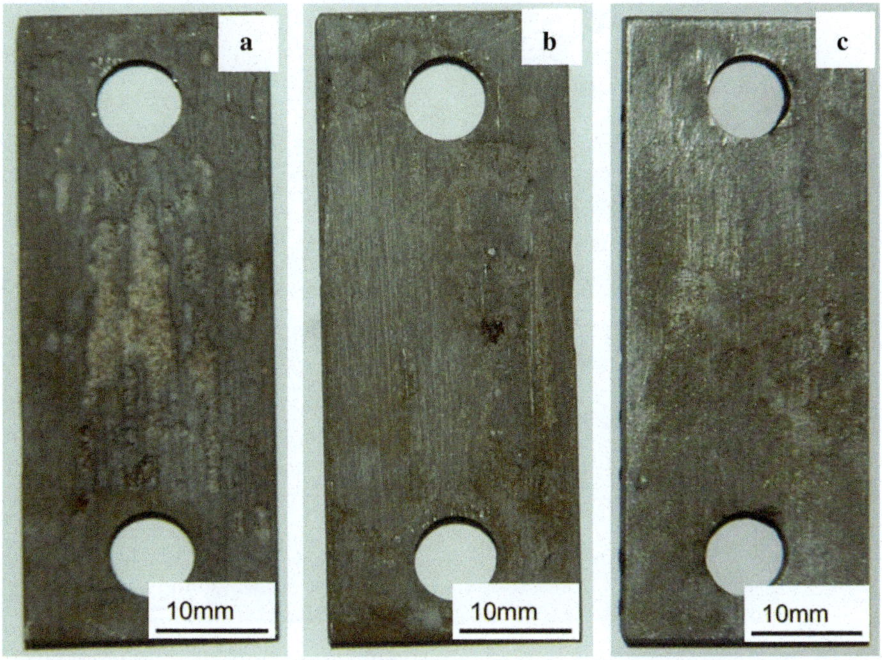

Fig. 5.1 Sample surface morphologies after corrosion scale removal in CO_2-saturated water for 240 h: **a** N80 steel with no Cr (tempered martensitic microstructure); **b** N80 steel with 1 wt% Cr (tempered martensitic microstructure) and **c** N80 steel with 1 wt% Cr (ferritic-pearlitic microstructure) [2]

titanium is also very stable at high temperatures. Therefore, for harsh environemnts with high temperature and high concentrations of SO_2 or H_2S, a small amount of titanium can be added to the alloy to achieve good corrosion-resisting performance.

Previous studies have focused on the corrosion behaviors of various Ni-base and Ni–Ti-base alloys in wet CO_2 and supercritical CO_2 environments at broad temperature and pressure ranges [8–12]. Generally, it has been reported that Ni-base and Ni–Ti-base alloys exhibited superior corrosion resistances, especially at high temperatures. For example, Xie [13] exposed Ni-25 wt% Cr and Ni-30 wt% Cr alloys to a flowing Ar-20 vol% CO_2-20 vol% H_2O mixture with a linear flow rate of 2 cm/s and a total pressure of 1 atm at 700 and 800 °C for up to 500 h. Both alloys formed a continuous chromia layer on their surfaces after 500 h of corrosion, resulting in much reduced corrosion rates. However, the Ni–Ti-base alloys can suffer noticeable degradation when the alloys are loaded with high tensile strength (where stress fatigue becomes significant), and the supercritical CO_2 exposure pressure becomes high. Kim [14] tested the corrosion-resisting performance of a Ni and Ti-bearing alloy (73.4 wt% Ni and 0.2 wt% Ti) in an extreme supercritical CO_2-rich environment ($T = 650$ °C and $P = 20$ MPa). All the samples were under > 130 MPa tensile stress during corrosion. They found that the applied tensile stress resulted in substantially

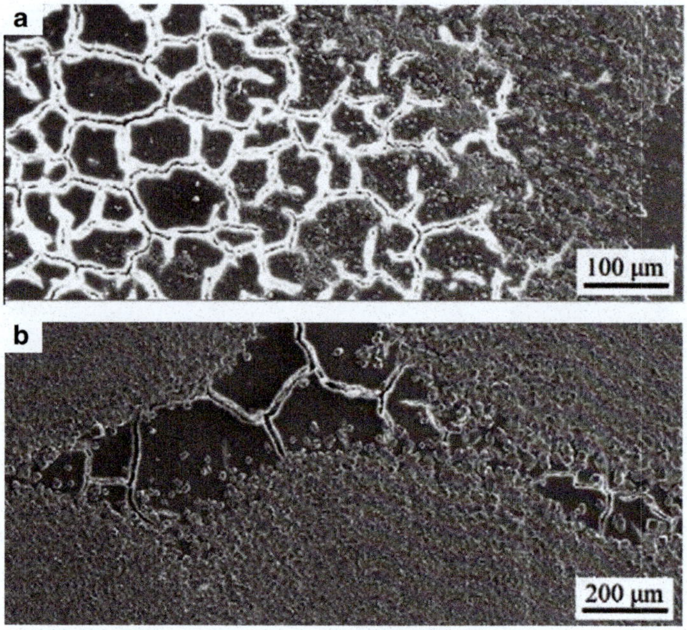

Fig. 5.2 Formation of Cr-bearing scale as a result of the reaction between Cr and CO_2. **a** After 32 h of reaction; **b** after 222 h of reaction [1]

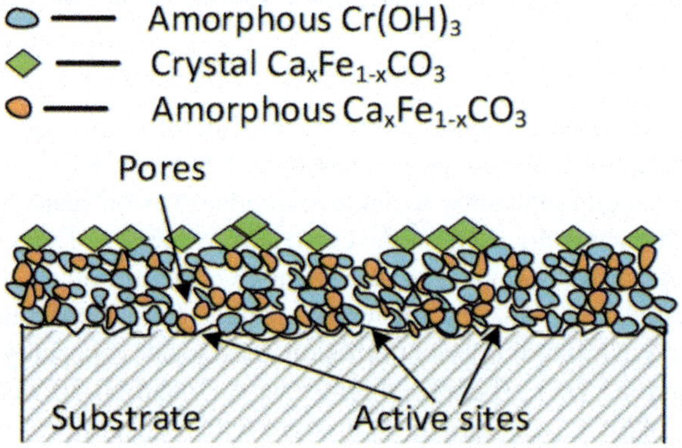

Fig. 5.3 Scheme of formation of $Cr(OH)_3$ and other scales (besides Cr-bearing scale and $FeCO_3$) as a result of the reaction between the alloy and CO_2 (modified from [6])

increased oxidation rates. Oxide layers formed during creep in supercritical CO_2 were thick due to combined effects of tensile stress, high supercritical CO_2 pressure, and high temperature. Based on the experimental results, Kim [14] reached a conclusion that the thick oxide layers formed on Ni–Ti-base alloys are very prone to cracking and spallation. When a crack occurs, the crack propagates through the oxide layer, which in turn results in further oxidation occurring around and ahead of the crack. In summary, the Ni–Ti-alloy has a superior corrosion resistance under normal wet CO_2 or supercritical CO_2 conditions. However, when the samples are loaded with high tensile stress and the supercritical CO_2 partial pressure reaches 20+ MPa, the tensile stress loading contributes to an obvious degradation of the steel alloy, even though the alloy contains Ni and Ti. A practical guide for optimal alloy selection has been provided by Rackley [15] (Table 5.1).

The CO_2-corrosion resistance of special alloys like Ce-bearing alloys and Nb-bearing alloys have also been tested. The corrosion behavior of Fe-15 wt% Ce alloy at 700 °C in H_2–CO_2, H_2–H_2S, and H_2–H_2S–CO_2 gas combinations was studied by Fu [16]. When the Fe-15 wt% Ce alloy corroded in H_2–CO_2 combinations, an internal oxidation in the Fe-15 wt% Ce alloy was observed. The intricate scales developed when the alloy corroded in combinations of H_2–H_2S and H_2–H_2S–CO_2. In any case, no exclusive Ce scales developed. This occurs as a result of the alloy's intermetallic complexes and the limited solubility of Ce in the base metal. At 700 °C, the corrosion rate of the alloy was lower in the H_2–H_2S–CO_2 mixture than it was in the H2-H_2S mixture, and the corrosion rate was lowest in the H_2–CO_2 mixture. Wang [17] tested the corrosion-resisting performance of a Nb- and Ti-bearing alloy (0.05 wt% Nb and 0.03 wt% Ti) when exposed to 3.5 wt% NaCl solution saturated by 0.64 MPa CO_2. Compared with the control group, the corrosion rate of the Nb- and Ti-bearing alloy was 8.4% lower at $t = 25$ h and 14.6% lower at $t = 380$ h, showing that the presence of Nb and Ti improved the corrosion resistance of the alloy.

In summary, Cr-bearing steel alloys are the most widely used CO_2-resisting alloys. 1–3 wt% Cr in alloys is sufficient to bear CO_2 corrosion under normal operating conditions. With high tensile strength loading and high partial pressure of CO_2, Ni-base and Ni–Ti-base alloys are recommended to resist CO_2 corrosion. In all scenarios,

Table 5.1 Advantages and disadvantages of various alloy types

Alloy type	Composition	Advantages	Disadvantages
Ferritic alloy	Fe–Cr alloy steels with 10–25% Cr, 2–4% Mo, < 1% Ni, and < 0.75% C	Lower thermal stresses under cyclic operation due to a lower thermal expansion coefficient, especially for thick-walled components	Low creep strength of traditional ferritic steels requires alloying such as tungsten and vanadium
Austenitic alloy	Fe–Cr-Ni alloys with 16–25% Cr, 1–37% Ni, and < 0.24% C	High corrosion resistance, which is highly desired in corrosive service such as refuse incineration boilers	Higher thermal expansion and low creep strength for high-temperature service
Martensitic alloy	Fe–Cr alloys with 12–18% Cr and < 1% C; may also include up to 2% Ni, Mo, V, or W	The microcrystalline structure contributes to a high creep resistance	Poorer corrosion resistance, which can be improved by precipitation hardening; more brittle

Note This table is modified from [15]

regular monitoring of the alloys is required to ensure that small pit corrosion can be detected at the early stage. It is very important to note that the selection of CO_2-resistant steel alloys for applications in corrosive CO_2 environment can be a complex procedure and if improperly carried out can lead to mistakes in application and misunderstanding about the performance of a CO_2-resistant steel in a specific service environment.

5.2 Corrosion-Resistant Cement

By preparing cement stone with high compactness, the micro- and nano-pores formed in the process of cement slurry hydration are reduced, and the permeability of cement stone is reduced, which can inhibit the corrosion process of high concentration CO_2. If the SBR latex is used in the cementing system, the pores in the cement stone can be reduced because the latex forms a film in the cement stone, thereby reducing the permeability of the cement stone, and also blocking the contact between the corrosion agent and the cement stone. By adding SiO_2 and Al_2O_3 to the cement slurry, the volcanic ash reaction occurs to generate C–S–H, which reduces the permeability of the cement stone and reduces the content of $Ca(OH)_2$ in the cement stone. Since C–S–H is a good crosslinking agent and $Ca(OH)_2$ is prone to CO_2 attack, adding SiO_2 and Al_2O_3 to the cement slurry can greatly improve the corrosion resistance of the cement stone.

A good compatibility with high temperature is the basis for the design of anti-corrosion cementing system, because it is common to have high temperatures in deep subsurface. The high-temperature stability of the cement depends on the contents of thermally stable admixtures in the cement. In order to ensure that the cement slurry has a good stability at high temperature, the research is recommended to focus on thermally-stable cement slurry admixture, settlement of cement slurry at high temperature, and strength evolution of cement stone at high temperature.

Cement slurry needs to have good anti-gas channeling ability, which is the focus of the design of high-temperature and high-pressure anti-gas channeling cement slurry system. There are four design ideas for the anti-channeling performance of cement slurry. The first is to reduce the permeability of the cement stone as much as possible. Reducing the permeability of the cement stone can inhibit gas permeation and delay gas intrusion into the cement. The second is to reduce the transition time from cement slurry to solid cement stone, and thus reduce the probability of gas channeling and shorten the migration time of gas in the cement slurry. The third is to use a cement slurry system with a low volume shrinkage rate to prevent the formation of microannular gaps. The fourth is to use high-toughness cement stone, which reduces the elastic modulus, improves the plastic limit strain rate, enhances the loading resistance of cement stone, and prevents the formation of microcracks.

One of the most effective cement slurry systems against CO_2 corrosion is the composite cement system with phosphoaluminates. Aluminate cement and phosphate cement have higher CO_2 corrosion resistance than ordinary Portland cement, because the minerals and chemical components of aluminate cement and phosphate cement are different from those of ordinary Portland cement. Aluminate cement and phosphate cement have improved structure and their hydration products are resistant to CO_2 attack, which reduces the corrosion rate of CO_2. Li [18] added phosphoaluminate cement as an admixture to ordinary Portland cement for research. The results show that a certain amount of phosphoaluminate cement can improve cement paste's mechanical strength and water resistance. According to the analysis, $(AlO_4)^{5-}$ can combine with calcium ions in ordinary Portland cement to form more stable hydration products. Ren [19] believe that after the composite cement of phosphoaluminate cement and portland cement is hydrated, due to the formation of gel materials such as C–A–P–H and C–P–H, the generated ettringite will hardly undergo phase transformation to monosulfide ettringite, which also ensures that the hydrated products of the composite cement after hydration are more stable and dense, and improves its impermeability. Ding [20] studied the micropore structure of phosphoaluminate cement and ordinary Portland cement using the isothermal adsorption method. It was found that after the hydration of phosphoaluminate cement, the large pore size was less and the total porosity was significantly lower than that of ordinary Portland cement. Therefore, the compact internal structure guarantees good impermeability of phosphoaluminate cement. Liu [21] studied the surface air permeability and water absorption of phosphoaluminate cement concrete, and the results showed that the surface air resistance and water absorption of phosphoaluminate cement concrete were better than ordinary Portland cement. SEM and MIP were used to verify that the pore structure of phosphoaluminate cement is excellent with few macropores. The dense structure of hydrated slurry is another proof of its good impermeability. Liu [22] believed that the hydration products of phosphoaluminate cement absorbed OH^- during the hydration process, which reduced the alkalinity of the hydration system and promoted the hydration of C_2S and C_3S. In addition, the distribution of the internal voids of the hardened composite cement paste is mainly concentrated in the pores, which significantly improves the overall performance of the composite cement-based material and the water resistance.

Li [23] studied the hydration activity of the aluminium-rich zone in the ternary system $CaO–Al_2O_3–P_2O_5$ and synthesized a new type of phosphoaluminate cement, which has different mineral phases from Portland cement. The main mineral phases are the LHss phase (a kind of phosphoaluminate solid solution), calcium aluminate (CA) and a small amount of glass. The main chemical components of phosphoaluminate cement (PALC) are Al_2O_3, CaO, P_2O_5, SiO_2, and a small amount of MgO, Fe_2O_3, and SO_3. The main hydration products of phosphoaluminate cement are hydrated phosphate gel (C–P–H), hydrated phosphoaluminate (C–A–S–H) and aluminium gel. As the hydration product of phosphoaluminate cement does not contain $Ca(OH)_2$, which is easily corroded by CO_2, the phosphoaluminate cement has great CO_2 corrosion resistance. Therefore, combining phosphoaluminate cement and Portland cement to prepare a phosphoaluminate-Portland composite cement system can improve the CO_2 corrosion resistance of cement [24]. The cement slurry formula of the composite cement system can be abbreviated as follows: G (Class G cement)

+ X% PALC (phosphoaluminate cement) + 3 ~ 5% micro silica + 1.5 ~ 2.5% G33S (fluid loss agent) + 1 ~ 3% SR (retarder) + water.

Currently, the CO_2 corrosion resistance of cement paste is usually based on the change in compressive strength and permeability of cement paste before and after CO_2 corrosion. Portland cement and phosphoaluminate cement are mixed with micro silica and additives, respectively, and then a cement slurry system with a density of 1.85 g/cm^3 is prepared according to a certain water-cement ratio. The cement slurry system is cured at 30 °C, 50 °C and 70 °C for 7 days, respectively. After forming, it is placed in a high-temperature, high-pressure kettle for corrosion tests. The corrosion conditions are: 90 °C, 120 °C, $P_{CO2} = 3$ MPa, $P_{N2} = 7$ MPa, and corrosion lasts 7 days. The compressive strength of cement paste of different systems before and after CO_2 corrosion is shown in Table 5.2. The test results show that the compressive strength of Portland cement declines after 7d of corrosion at 120 °C, $P_{CO2} = 3$ MPa, $P_{N2} = 7$ MPa.

For phosphate cement, Table 5.2 shows that the compressive strength of cement paste cured at 30 °C decreases after being corroded by carbon dioxide, and its compressive strength attenuation rate is 10%. The strength of cement paste after corrosion is 21.38 MPa, still more significant than 14 MPa. The compressive strength of cement paste cured at 50 and 70 °C increases after carbon dioxide corrosion. The strength of cement paste cured at 50 °C after carbon dioxide corrosion reaches 41 MPa, and the performance of cement paste after corrosion is better than that before corrosion. The compressive strength growth rate of cement paste cured at 70 °C after

Table 5.2 The comparison of the strength of cement paste of different systems before and after CO_2 corrosion

Cement system	Curing temperature/°C	Corrosion temperature/°C	Corrosion pressure/MPa	Before corrosion	After corrosion	
				7d strength/MPa	7d strength/MPa	Attenuation rate/%
Class G cement	30	90	$P_{CO2} = 3$ $P_{N2} = 7$	18.4	9.8	46.8
	50			22.2	12.6	43.4
	70			18.3	10.5	42.1
	30	120		18.4	14.7	20.1
	50			22.2	13.5	39
	70			18.3	12.3	32
Phosphoaluminate cement	30	90		23.8	21.4	10
	50			22.4	41.2	– 83.6
	70			15.5	17.0	– 9.6
	30	120		23.8		
	50			16.7	18.2	– 8.8
	70			26.8	32.8	– 22.3

carbon dioxide corrosion is 9.6%, and the strength is 17 MPa. Phosphoaluminate cement has excellent corrosion resistance to carbon dioxide.

Adding PALC to ordinary Portland cement can also enhance the CO_2 corrosion resistance. Table 5.3 shows the compressive strengths of G Class + 40%/50% PALC samples after reacting with CO_2 at 30, 50 and 70 °C. The results show that the composite cement systems have good CO_2 corrosion resistance at 70 °C. The strength change of samples cured at 30 and 50 °C is not uniform before and after corrosion due to the complex composition of the composite cement system. However, the experimental results show that the attenuation rate of G + 5% PALC and G + 3% PALC samples is low at 30 and 50 °C. Therefore, under low temperatures, the content of phosphoaluminate in a composite cement system should not be too high.

Table 5.4 compares the porosity and permeability of each group of the G + PALC samples before and after CO_2 corrosion after curing at 50 °C for 7d. The results show that the PALC cement has excellent corrosion resistance, and the G Class + 3%/5% PALC composite cement system has strong corrosion resistance after curing at 50 °C for 7 days.

The composite cement system sample with better corrosion resistance cured at 70 °C for 7d was corroded for one month under the corrosion conditions of 120 °C, $P_{CO2} = 3$ MPa and $P_{N2} = 7$ MPa to evaluate its long-term durability. The experimental results are shown in Fig. 5.4.

As shown in Fig. 5.4, the compressive strength of the Class G Portland cement sample sharply declined after CO_2 corroded the sample for 1 month. The strength of the G + 50% PALC composite cement sample also had attenuation, while the compressive strengths of phosphoaluminate cement and G + 40% PALC composite cement system increased under the same conditions. Therefore, to meet the integrity requirements of the cement sheath, both the PALC cement and the G + 40% PALC composite cement systems had good CO_2 corrosion resistance under long-term corrosion conditions.

For the phosphoaluminate cement system, the main minerals after curing at 50 °C for 7 days are hydrated calcium phosphate ($Ca_5P_6O_{20}{\cdot}H_2O$), aluminium hydroxide ($Al(OH)_3$), hydrated calcium phosphoaluminate ($Ca_2Al_2P_4O_{15}{\cdot}H_2O$), and hydrated calcium aluminosilicate ($Ca_2Al_2SiO_7{\cdot}8H_2O$). After the phosphoaluminate cement system is corroded by CO_2, the main mineral components of the first reaction layer is $Al(OH)_3$, $Ca_5P_6O_{20}{\cdot}H_2O$, $Ca_{10}(PO_4)(CO_3)_3(OH))$, $CaPO_3(OH){\cdot}2H_2O$, and $Ca_2Al_2SiO_7{\cdot}8H_2O$. The main mineral components of the second reaction layer are $Al(OH)_3$, $Ca_5P_6O_{20}{\cdot}H_2O$, $Ca_{10}(PO_4)(CO_3)_3(OH)_2$, and $Ca_2Al_2SiO_7{\cdot}8H_2O$. The main mineral components of the third and the fourth reaction layers are $Al(OH)_3$, $Ca_5P_6O_{20}{\cdot}H_2O$, and $Ca_2Al_2SiO_7{\cdot}8H_2O$. According to the above phase analysis results, there is no $Ca(OH)_2$ in the phosphoaluminate cement system, so there is no $CaCO_3$ formation after CO_2 corrosion. Compared with $CaCO_3$, the main carbonate formed in phosphoaluminate cement (i.e., $Ca_{10}(PO_4)(CO_3)_3(OH)_2$) has a higher CO_2 resistance. $Ca_{10}(PO_4)(CO_3)_3(OH)_2$ in reaction layers 1 and 2 can slow down the corrosion of CO_2 in the phosphoaluminate cement system, and layers 1 and 2 contain $Al(OH)_3$, $Ca_5P_6O_{20}{\cdot}H_2O$ and other components that do not react with CO_2, indicating that the phosphoaluminate cement system has good CO_2 corrosion resistance.

Table 5.3 The change of compressive strength of composite cement system after corrosion

G + X%PALC	BC (30 °C)/MPa	AC (30 °C)/MPa	AR/%	BC (50 °C)/MPa	AC (50 °C)/MPa	AR/%	BC (70 °C)/MPa	AC (70 °C)/MPa	AR/%
0%	18.4	9.8	46.8	22.2	12.6	43.4	18.2	10.5	42.1
3%	19.5	18.6	4.2	24.8	22.3	10.1	22.2	20.4	8.1
5%	22.3	31.8	−42.3	23.2	16.0	30.8	22.0	17.2	22.0
10%	18.3	15.0	18.4	23.0	12.3	46.5	22.7	18.9	16.8
15%	18.2	14.7	19.1	22.2	13.1	41.0	20.4	16.9	17.5
20%	17.2	13.7	20.2	23.2	13.7	41.0	21.5	15.3	28.2
30%	17.1	8.1	52.4	24.4	11.8	51.7	22.3	12.2	45.2
40%	16.1	9.1	43.4	23.3	12.8	45.0	21.8	49.7	−128.0
50%	14.2	7.2	49.6	22.6	12.1	46.2	13.3	22.5	−69.1
60%	12.4	7.4	39.9	15.0	7.9	47.4	12.0	9.2	23.2
70%	13.7	7.4	46.1	13.6	7.4	45.4	14.2	16.7	−17.4
100%	10.5	3.3	68.7	18.2	7.4	59.5	14.3	9.5	33.6
PALC	23.8	21.4	10.0	22.4	41.2	−83.6	15.5	17.0	−9.6

BC Before Corrosion, *AC* After Corrosion, *AR* Attenuation Rate

Table 5.4 The comparison of porosity and permeability before and after corrosion

Cement system	Before corrosion		After corrosion	
	Porosity (%)	Permeability (mD)	Porosity (%)	Permeability (mD)
G + 3%PALC	29.6	0.021	33.96	0.033
G + 5%PALC	29.4	0.025	34.25	0.041
G + 30%PALC	31.8	0.032	46.4	0.047
G + 50%PALC	30.2	0.026	51.37	0.042
G + PALC (1:1)	39.43	0.041	42.6	0.055
PALC	30.1	0.043	12.24	0.014
G	30.2	0.046	51.52	0.060

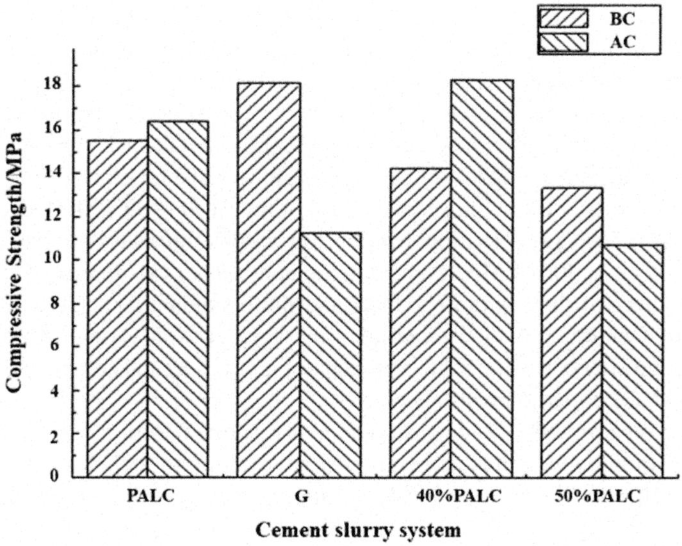

Fig. 5.4 The comparison of compressive strength before and after corrosion

After the G + 40% PALC composite system cement paste is corroded by CO_2, the phase composition of each layer in the sample is as follows: the main mineral components of the first layer are $CaCO_3$, hydrated calcium silicate gel ($Ca_{1.5}SiO_{3.5}\cdot xH_2O$), $Ca_5P_6O_{20}\cdot H_2O$ and $Al(OH)_3$; The main mineral components of the second to fourth layers are $Al(OH)_3$, $Ca_{1.5}SiO_{3.5}\cdot xH_2O$, $CaSiO_4\cdot H_2O$ and $Ca_5P_6O_{20}\cdot H_2O$. Therefore, only a small amount of $CaCO_3$ is generated on the surface of G + 40% PALC composite cement paste after CO_2 corrosion, and the surface also contains $Ca_{1.5}SiO_{3.5}\cdot xH_2O$, $Ca_5P_6O_{20}\cdot H_2O$ and $Al(OH)_3$. These components form a protective barrier to the cement paste, prevent the corrosion invasion of CO_2, and give it excellent corrosion resistance.

The main mineral compositions of the cement paste cured at 70 °C for 7 days in the G + PALC composite system (1:1) are ($Ca_3Al_2(SiO_4)_{1.25}(OH)_7$), $Ca(OH)_2$, $Ca_{1.5}SiO_{3.5}\cdot xH_2O$ and calcium superphosphate ($Ca_2P_2O_7$). The products of each layer of the sample after CO_2 corrosion are as follows: the main mineral components of the first to second layers are $CaCO_3$ and $Ca_2P_2O_7$; The main mineral components of the third layer are $CaCO_3$, $Ca_2P_2O_7$, $Ca_2Al_2SiO_7\cdot 8H_2O$ and $Ca_3(PO_4)_2$. Therefore, after the G + PALC composite system (1:1) is corroded by CO_2, $CaCO_3$ is generated from inside to outside, and the content is high, indicating the corrosion resistance of the G + PALC composite system (1:1) is poor.

The microscopic morphology of the corrosion products on the surface of Class G Portland cement after being corroded by CO_2 is mainly needle-like and blocky, which show typical $CaCO_3$ crystal morphology. Therefore, it is verified that a large amount of $CaCO_3$ is generated on the surface of Class G Portland cement after corrosion. Figure 5.5 shows the microscopic morphology of phosphoaluminate (PALC) cement after CO_2 corrosion. Its outer layer structure is relatively compact, and different morphologies and structures are interspersed, which is conducive to protecting its internal structure from corrosion. In the samples' internal structure, the products are closely connected, with prominent crystal structures, and the acicular blocky crystal forms interlaced. The compact structure can help to resist CO_2 corrosion.

Figure 5.6 shows the microscopic morphology of G + 40% PALC composite cement paste after CO_2 corrosion. Similar to phosphoaluminate cement, after CO_2 corrosion of G + 40% PALC composite cement paste, the microstructure of the outermost layer of the sample is dense, and the crystals are staggered and cross-linked with each other. No loose particles and apparent $CaCO_3$ morphology are on the outer edge after corrosion. The cement paste is compact and complete. Therefore, it has good CO_2 corrosion resistance.

Another type of anti-CO_2 corrosion cement is cement with inorganic or organic additives. Common inorganic anti-CO_2 corrosion additives include nanoclays, nano silica, Fe_2O_3, etc. Mei [25] proposed an approach to modify Ca-montmorillonite

Fig. 5.5 Micromorphology of phosphoaluminate cement after CO_2 corrosion

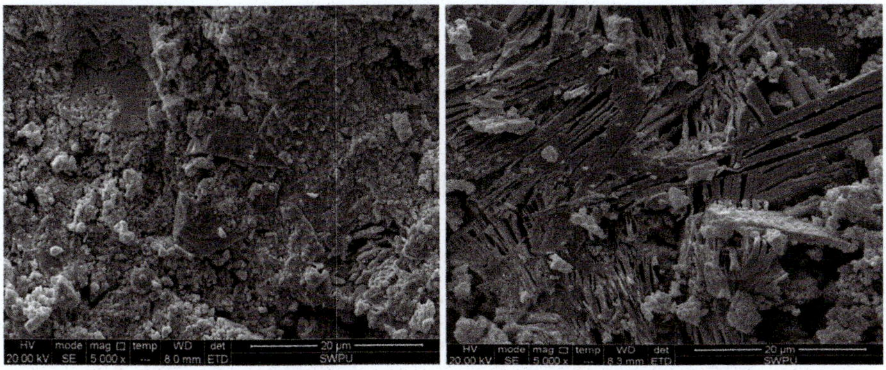

Fig. 5.6 Micro morphology of G + 40% PALC composite cement paste after CO_2 corrosion

(Ca-MMT) nanoclays using supercritical CO_2 ($ScCO_2$) as the solvent and intercalator. The Ca-MMT nanoclays were immersed in $ScCO_2$ at a pressure of 8.0 MPa and temperature of 60 °C in a high-pressure reactor for 24 h. After $ScCO_2$ modification, the nanoclays were oven-dried in an isolated environment to stabilize the composition and structure of the CO_2-modified Ca-MMT nanoclays. The CO_2-modified Ca-MMT nanoclays were then added to the cement slurry to reinforce the oilwell cement. A CO_2 corrosion experiment of wellbore cement with CO_2-modified Ca-MMT nanoclay was operated in a high temperature and high pressure (HTHP) kettle, achieving a high concentration CO_2 corrosion condition. Micro-CT assessment of wellbore cement with CO_2-modified Ca-MMT nanoclay showed that the expansion of heavy carbonation areas was restricted after 28 days CO_2 corrosion (Fig. 5.7). In summary, the modified nanoclay prevents the permeation of CO_2 into the wellbore cement and seals the corroded area, which improves the durability of wellbore cement. Mostafa [26] found that adding nano silica into concrete and cement could effectively improve the microstructure, strength and durability of concrete and cement, which plays a vital role in lowering the diffusion of corrosion agents in concrete and cement. Xu [27] investigated the corrosion-resistant properties of corrosion-resistant additive for Fe_2O_3-amended cement. They found that Fe_2O_3 could react with $Ca(OH)_2$ and high-Ca/Si hydration products to generate low-Ca/Si hydration products such as xonotlite and tobermorite. The Fe_2O_3-amended cement had superior corrosion resistance because of its low original permeability due to film formation and filling effects and through the reduction of $Ca(OH)_2$ to achieve low-Ca/Si hydration products.

The most common organic anti-CO_2 corrosion cement additive is epoxy resin. Peng [28] improved the anti-corrosion performance of oil well cement by adding water-based modified epoxy resin as a cement additive. They discovered that the cement with modified epoxy resin greatly increased its corrosion resistance. The corrosion depths were 3.1 mm and 10.6 mm, respectively, after 30 days of corrosion in liquid and gas phase acidic medium, with strength loss rates of 9.8% and 1.2%, respectively. Under the same corrosion conditions, the corrosion depths were 16.1 mm and 36.6 mm, and the strength loss rates were as high as 32.1% and 17.6%

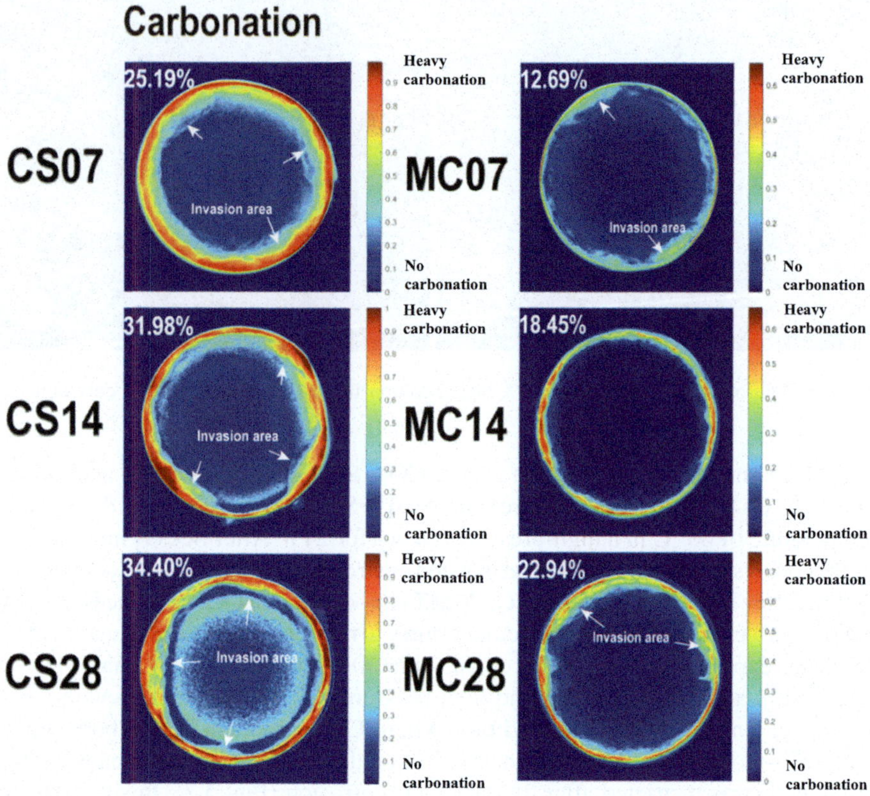

Fig. 5.7 Distribution maps of carbonates in the control group (CS) and CO_2-modified Ca-MMT group (MC) during 7, 14 and 28 days CO_2 corrosion (Red: heavy carbonation; Blue: slight carbonation) [25]

compared to the cement without modified epoxy resin. Another common organic anti-CO_2 corrosion cement additive is latex. Yang [29] mixed latex, superfine pitch, and noncrystalline SiO_2 to produce a new carbonation resistant additive to improve carbonation resistance properties of oil well cement by decreasing original permeability, $Ca(OH)_2$ and calcium to silicon ratio (C/S). They found that decreasing C/S was an effective measure for improving cement resistance to carbonation, and the additive resulted in a remarkable decreasing in $Ca(OH)_2$, which was changed into carbonation resist products (C–S–H). As a result, the C/S was decreased. Upon mixing the cement with the carbonation resistant additive, the original permeability and carbonation depth decreased, and the compressive strength increased. The additive weakened the leaching and dissolution effect by consuming $Ca(OH)_2$ and high C/S hydration products.

5.3 On-Site Practice

China's first CCS CO_2 storage site, Shenhua CCS site, applied S13 Cr as the building material for the oil tubes of the CO_2 injection well and the first monitoring well. The oil tube of the CO_2 injection well had an external diameter of 73 mm and an inner diameter of 62 mm. The oil tube of the first monitoring well had an external diameter of 73 mm and an inner diameter of 62 mm, which were the same as those of the CO_2 injection well. The S13 Cr was used as the oil tube building material based on the consideration that supercritical CO_2 was in direct contact with the oil tube, and a material with superior CO_2 corrosion resistance such as S13 Cr was preferred. Well logging confirmed that the tubing of the CO_2 injection well and the monitoring well maintained the integrity and suffered no corrosion by CO_2. Shenhua Ordos CCS project is the first whole-chain CCS demonstration project in China, which injects CO_2 into the saline aquifer. The CO_2 injection started in October 2010 and ceased in April 2015 with a total amount of 300,000 tons CO_2 injection. The injected CO_2 came from the high concentrated CO_2 gas in the first worldwide large-scale direct coal-to-liquid plant after liquefaction, purification and transportation. An aerial view of the Shenhua CCS demonstration site and its surroundings is shown in Fig. 5.8. CO_2 was compressed to about 20 MPa through the pump in order to reach the pressure of the saline reservoir in the depth of ~ 1699 m. CO_2 is also required to heat before injection in some cases for preventing the possible blockage. The CO_2 stream was with 99.9% purity, and more than 300,000 tons CO_2 was cumulatively injected into the saline aquifer [30].

For the CO_2 injection well and the first monitoring well, S13 Cr was the building material for the well casing. Corrosion tests showed that the maximum corrosion rate of the S13 Cr casing exposed to CO_2 and water given the subsurface pressure and temperature was 0.1 mm/year. Given a casing thickness of 5.5 mm, the worst-case breakthrough time was 55 years. However, given the protection of the surrounding cement ring, the breakthrough time was expected to be much longer than 55 years. For the N80 casing, the maximum corrosion rate of the N80 casing exposed to CO_2 and water given the subsurface pressure and temperature was 5 mm/year. Given a casing thickness of 5.5 mm, the worst-case breakthrough time was 1.1 years. Therefore, for the N80 casing, noticeable CO_2-induced corrosion was possible since the corrosion resistance of N80 casing was much lower than that of S13 Cr casing.

The Shenhua CCS site has set up strict health, safety and environmental (HSE) management rules. The site requires the construction operators to strictly follow the laws and regulations set up by central government, local government, and oil and gas industry. The construction operators shall set up a HSE management framework and conduct regional safety and environmental risk analysis of the site. Measures for blowout risk minimization and emergency responding plans for blowout accidents shall be prepared, and regular drills for emergency response shall be conducted. For safety signs, appropriate warning signs shall be placed at the well drilling platform, pump room, compressor room, oil tanks, gas storage tanks, etc. Clear signs shall be placed at the routes for emergency evacuation. Signs for emergency gathering

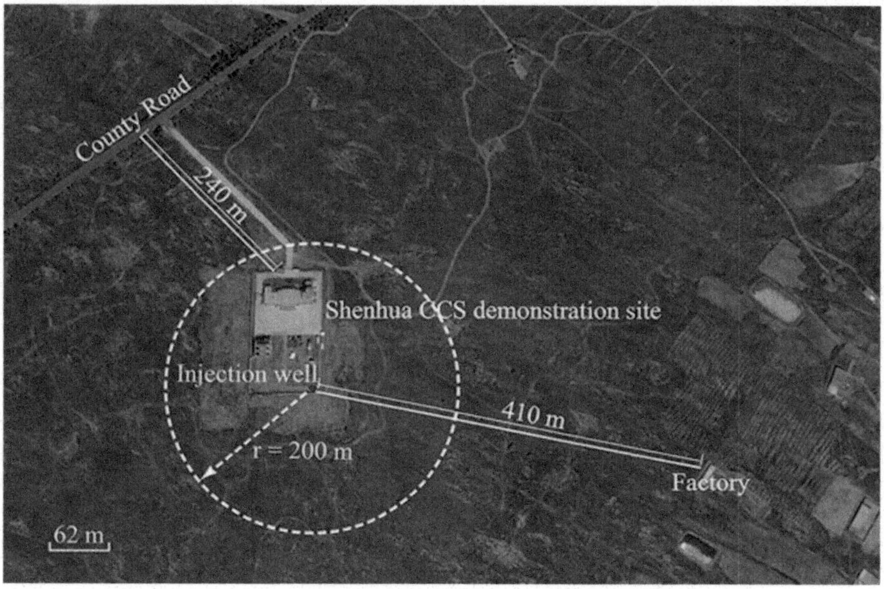

Fig. 5.8 Aerial view of the Shenhua CCS demonstration site and its surroundings [31]

locations and waste disposal locations shall be placed as well. The site shall prepare enough fire fighting equipment in accordance with fire safety regulations, and a special person shall be assigned to be responsible for regular inspection of fire fighting equipment. Fire extinguishers shall be regularly checked to avoid malfunction, and the manufacturing date and the inspection card signed by the inspector shall be placed on the fire extinguishers. Pipelines, valves, steel tanks, etc. on site shall be regularly checked for signs of corrosion.

References

1. Guo, S., Xu, L., Zhang, L., Chang, W., Lu, M.: Corrosion of alloy steels containing 2% chromium in CO_2 environments. Corros. Sci. **63**, 246–258 (2012)
2. Wu, Q., Zhang, Z., Dong, X., Yang, J.: Corrosion behavior of low-alloy steel containing 1% chromium in CO_2 environments. Corros. Sci. **75**, 400–408 (2013)
3. Sun, J., Sun, C., Lin, X., Cheng, X., Liu, H.: Effect of chromium on corrosion behavior of P110 steels in CO_2-H_2S environment with high pressure and high temperature. Materials **9**(3), 200 (2016)
4. Kermani, B., Gonzales, J.C., Turconi, G.L., Perez, T.E., Morales, C.: In-field corrosion performance of 3%Cr steels in sweet and sour downhole production and water injection, Corrosion. OnePetro (2004)
5. Sun, J., Liu, W., Chang, W., Zhang, Z., Li, Z., Yu, T., Lu, M.: Characteristics and formation mechanism of corrosion scales on low-chromium X65 steels in CO_2 environment. Acta Metall Sin **45**(1), 84–90 (2009)

6. Wang, B., Xu, L., Liu, G., Lu, M.: Corrosion behavior and mechanism of 3Cr steel in CO_2 environment with various Ca^{2+} concentration. Corros. Sci. **136**, 210–220 (2018)
7. Craig, B.D., Smith, L.: Corrosion resistant alloys (CRAs) in the oil and gas industry. Nickel Institute Technical Series **1**, 0073 (2011)
8. Firouzdor, V., Sridharan, K., Cao, G., Anderson, M., Allen, T.R.: Corrosion of a stainless steel and nickel-based alloys in high temperature supercritical carbon dioxide environment. Corros. Sci. **69**, 281–291 (2013)
9. Lee, H.J., Kim, H., Kim, S.H., Jang, C.: Corrosion and carburization behavior of chromia-forming heat resistant alloys in a high-temperature supercritical-carbon dioxide environment. Corros. Sci. **99**, 227–239 (2015)
10. Rouillard, F., Furukawa, T.: Corrosion of 9–12Cr ferritic–martensitic steels in high-temperature CO_2. Corros. Sci. **105**, 120–132 (2016)
11. Chen, H., Kim, S.H., Kim, C., Chen, J., Jang, C.: Corrosion behaviors of four stainless steels with similar chromium content in supercritical carbon dioxide environment at 650 °C. Corros. Sci. **156**, 16–31 (2019)
12. Gui, Y., Liang, Z., Zhao, Q.: Corrosion and carburization behavior of heat-resistant steels in a high-temperature supercritical carbon dioxide environment. Oxid. Met. **92**(1), 123–136 (2019)
13. Xie, Y., Nguyen, T.D., Zhang, J., Young, D.J.: Corrosion behaviour of Ni-Cr alloys in wet CO_2 atmosphere at 700 and 800 °C. Corros. Sci. **146**, 28–43 (2019)
14. Kim, S.H., Cha, J.-H., Jang, C.: Corrosion and creep behavior of a Ni-base alloy in supercritical-carbon dioxide environment at 650 °C. Corros. Sci. **174**, 108843 (2020)
15. Rackley, S.: Carbon capture from industrial processes, pp. 95–102. Butterworth-Heinemann, Oxford (2010)
16. Fu, G., Niu, Y.: Corrosion of Fe-15Ce alloy in three mixed-gas atmospheres. Trans. Nonferrous Met. Soc. China **12**(4), 874–877 (2002)
17. Wang, H., Wang, H., Gao, X., Yu, C.: Effect of Nb and Ti on corrosion characteristics of low alloy steel in supercritical CO_2 environment. Int. J. Electrochem. Sci. **14**, 10907–10919 (2019)
18. Li, S., Liu, B., Yi, Z., Ren, S., Zhai, G., Shu, L., Hu, J.: Water-resistance behavior of modified Portland cement mortar. J. Chin. Ceram. Soc. **33**, 1243–1247 (2005)
19. Ren, S., Zhai, G., Li, S., Hu, J., Zhang, N.: Study on water-resistance and mechanism of composite cement. Journal of University of Jinan (Sci. & Tech.) **3**, 207–211 (2003)
20. Zhu, D., Xiaodong, W., Ning, Z., Biao, L.: Paste strength and hydrates morphology of phosphoaluminate cement. J. Chin. Electron Microsc. Soc. **34**(06), 453–458 (2015)
21. Liu, P., Yu, Z., Huang, X., Zhang, X.: Research on the property of air permeability and sorptivity of phosphoaluminate cement concrete. J. Xi'an Uni.of Arch.& Tech.(Natural Science Edition) **42**(02), 216–220 (2010)
22. Liu, B., Zhai, G., Li, S., Zhang, N., Wang, W., Hu, J.: Research on hydration kinetics of composite cement of phosphoaluminate and silicate. J. Build. Mater. **3**, 259–265 (2008)
23. Li, S., Zhang, G., Zhang, N., Liu, B., Cao, W., Hu, J.: Study on hydraulic activity of aluminum-rich area in $CaO\text{-}Al_2O_3\text{-}P_2O_5$ system. J. Chin. Ceram. Soc. **2**, 16–23 (1998)
24. Wang, Y.: Study on the System of Phosphoaluminate and Silicate Composite Cement. Southwest Petroleum University, Chengdu (2014)
25. Mei, K., Zhang, L., Wang, Y., Cheng, X., Xue, Q., Gan, M., Fu, X., Zhang, C., Li, X.: Structural evolution in micro-calcite bearing Ca-montmorillonite reinforced oilwell cement during CO_2 invasion. Constr. Build. Mater. **315**, 125744 (2022)
26. Mostafa, S.A., El-Deeb, M.M., Farghali, A.A., Faried, A.S.: Evaluation of the nano silica and nano waste materials on the corrosion protection of high strength steel embedded in ultra-high performance concrete. Sci. Rep. **11**(1), 2617 (2021)
27. Xu, B., Yuan, B., Wang, Y.: Anti-corrosion cement for sour gas ($H_2S\text{-}CO_2$) storage and production of HTHP deep wells. Appl. Geochem. **96**, 155–163 (2018)
28. Peng, Z., Lv, F., Feng, Q., Zheng, Y.: Enhancing the $CO_2\text{-}H_2S$ corrosion resistance of oil well cement with a modified epoxy resin. Constr. Build. Mater. **326**, 126854 (2022)
29. Yang, Y., Yuan, B., Wang, Y., Zhang, S., Zhu, L.: Carbonation resistance cement for CO_2 storage and injection wells. J. Pet. Sci. Eng. **146**, 883–889 (2016)

30. Zhao, X., Ma, R., Zhang, F., Zhong, Z., Wang, B., Wang, Y., Li, Y., Weng, L.: The latest monitoring progress for Shenhua CO_2 storage project in China. Int. J. Greenhouse Gas Control **60**, 199–206 (2017)
31. Li, Q., Shi, H., Yang, D., Wei, X.: Modeling the key factors that could influence the diffusion of CO_2 from a wellbore blowout in the Ordos Basin, China. Environ. Sci. Pollut. Res. **24**(4), 3727–3738 (2017)

Chapter 6
Corrosion Control (II): Anti-corrosion Coating

Chenyang Deng and Liwei Zhang

6.1 General Introduction

Coating is a traditional anti-corrosion method. It "physically" blocks the interaction between CO_2 and the protecting target, like steel. Coatings can be applied on the internal and external surfaces of steel pipelines to protect them from corrosion induced by CO_2, other gases, and liquids. Temperature and humidity are the key factors to determine the appropriate type of coatings applied. One of the most popular anticorrosion techniques is the application of coatings on metals, which act as a barrier between the metal and its surroundings to prevent attack from corrosive agents. To locate and create coatings that are stronger and more durable, a lot of studies have been done [1]. Overall, the coating should be highly adhesive, resistant to corrosion and moisture, compatible to environment, and have enough structure strength to prevent any type of disbandment. High electrical insulation and enough electrical resistivity is required on coatings, and they should also withstand typical storage, handling and degradation. For applications where physical damage to the coating may occur, the coatings must be uniform, well adherent, pore free, and self-healing in order to provide effective corrosion protection [2].

Generally, the coating can be classified as organic coating and inorganic coating. The inorganic coating includes metal coating and non-metal coating. The most developed and widely applied organic coating is epoxy and its modified products. Besides, polyethylene is a traditional coating which was first applied in 1950s but is still being used at present. Recently, many new types of coating are undergoing research and development, such as graphene and polysiloxane. Meanwhile, some conventional coatings (e.g., asphalt, coal tar enamel, etc.) are gradually replaced by new types of

C. Deng · L. Zhang (✉)
State Key Laboratory of Geomechanics and Geotechnical Engineering, Institute of Rock and Soil Mechanics, Chinese Academy of Sciences, Wuhan 430071, China
e-mail: lwzhang@whrsm.ac.cn

University of Chinese Academy of Sciences, Beijing 100049, China

L. Zhang (ed.), *Corrosion in CO_2 Capture, Transportation, Geological Utilization and Storage*, Engineering Materials, https://doi.org/10.1007/978-981-99-2392-2_6

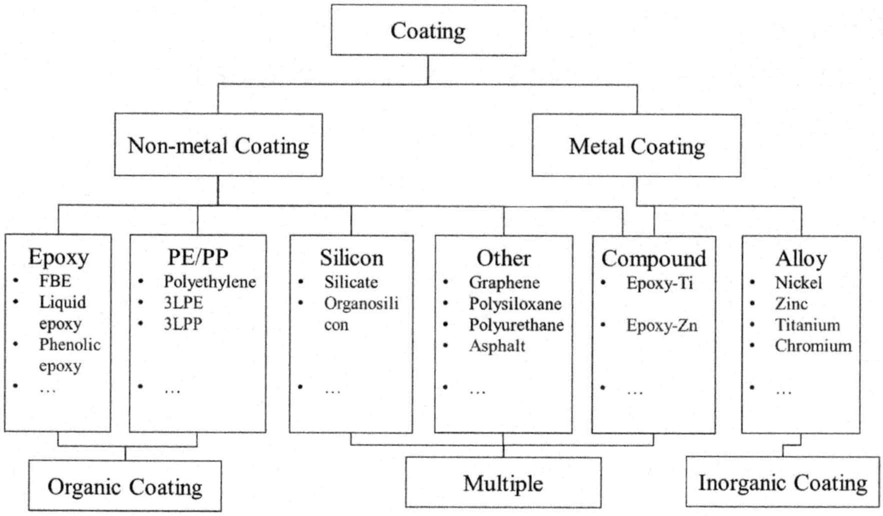

Fig. 6.1 Classification of coatings

coating due to the critical drawbacks of the conventional coatings. Inorganic coating includes silicon coating (some also belong to organic coatings as organosilicon) and a series of alloys such as nickel alloy, chromium alloy, and titanium alloy. In addition, the combination of metal and non-metal coating is a hot area where the related technology is rapidly developing (Fig. 6.1).

6.2 Representative Anti-corrosion Coatings

6.2.1 Epoxy Coating

Epoxy is the most economical and practical technic which has been widely applied in the industry. The epoxy has quite a number of excellent behaviors including: (1) wide range of varieties. With proper selection of resin and curing agents, epoxy could be produced ranging from low viscosity liquids to high melting point solid; (2) high adhesion. Due to the active surface functionalities on peripheral hydroxyl (–OH) groups and ether (–O–) bonds, epoxy exhibit strong adhesion ability with metallic surfaces; (3) low shrinkage. The common epoxy have shrinkage rate less than 2%; (4) mechanical properties. Cured epoxies possess high mechanical strength which ensure its industrial application; (5) electric properties. The epoxies have high electrical insulation and good electrical resistance at a wide range of temperature; (6) chemical stability. Cured epoxies act excellent chemical stability for acidic environment. (7) microbial resistance. The epoxies are strongly resistant to microbials, especially

Fig. 6.2 FBE coating on oil and gas pipelines (*Source* https://www.tjxysteel.com/oil-and-gas-line-pipe/fbe-coated-oil-and-gas-pipe-241.html)

against fungal. All of the characteristics of epoxy indicate it as the most preferable material on anti-corrosion coating [3].

Original epoxy coating has critical defects on structure strength. Its inner structure during cross-linking curing process exhibits fragility to some extent, and the curing process always creates plenty of micropores on the surface, impacting its structure integrity. Nowadays the manufacturers typically use additive or update the technology to produce modified epoxy products including fusion bonded epoxies (FBE), liquid epoxies and phenolic epoxies [3, 4]. The modified coatings may have several functions in addition to corrosion protection.

FBE coating was first introduced in the 1960s as a thermoset polymer coating. It is now the most commonly used pipeline epoxy coating in oil and gas industry (Fig. 6.2). Fusion bonding refers to a specific chemical cross-linking in which the dry FBE powder melts and forms liquid, then the liquid film wets and flows onto the steel surface, becoming a solid coating. This process is irreversible, once it takes place, the coating cannot be returned to the original state.

6.2.2 PE/PP Coating

Polyethylene/polypropylene (PE/PP) coating is one of the most common anti-corrosion coatings in the industry [5]. The 3-layer polyethylene or polypropylene (3LPE/3LPP) coating is a representative PE/PP coating. Technically, 3LPE/3LPP coating is an upgrade of FBE coating, due to that the FBE layer provides the corrosion resistance as part of the 3LPE/3LPP coating system. 3LPE/3LPP coating is generally applied on oil and gas pipelines and being increasingly used for water projects and CCUS projects as well. The excellent resistance of the 3LPE/3LPP coating to aging and erosion ensures its reliable protection [6]. It is composed of 3 layers: the first layer is a FBE coating acting as an anti-corrosion layer; the second layer is the adhesive layer which provides the adhesion between the FBE layer and the PE/PP layer; the third layer is the PE layer which provide not only the mechanical protection but also good resistance to heat and UV rays.

The 3LPE/3LPP coating is a versatile coating that can be applied to a wide range of pipeline diameters (from ~ 50 mm diameter to ~ 1220 mm diameter). The minimum recommended handling and construction temperature for the 3LPE/3LPP coating is − 40 °C, and the maximum recommended temperature is 85 °C. Compared with the FBE coating, the 3LPE/3LPP coating has better mechanical strength and higher resistance to erosion, but the cost is also high. In addition, it has more strict demand on the coating thickness and degradation of any layer may weaken the anti-corrosion effect.

6.2.3 Polyurethane Coating

Polyurethane (PU) coating is a polyurethane layer applied to the surface of a substrate, which has the similar protecting function with epoxy coating. PU is a type of polymer connected to a chemical compound group named carbamates. Same as epoxy, PU is also thermosetting in nature, meaning that it could not be melted when healed. Another characteristic of PU coating is that it could be formulated in varied gloss and transparency due to its customizability.

Although PU coating may look similar to epoxy coating, they possess several distinct properties that could be applied to different situations [7]: (1) PU coatings are relatively durable, softer and more elastic than epoxy ones, allowing them to absorb sharp impact loading; (2) PU coatings are more resistant to abrasion such as dents and scratches; (3) at low temperatures (lower than 0 °C), PU are more likely to sustain their shape and mechanical property than epoxy due to its improved elasticity. It has significant ability to adhere well to varieties of substrates in a broad range of temperature; (4) PU coatings have relatively shorter curing time, which ensure them to be put into use in 24 h while the epoxy coating needs at least 7 days for curing; (5) PU is not less resistant to wide a range of corrosive chemicals and it is more sensitive to ultraviolent, therefore it is not suitable to be directly exposed to UV light sources.

One notable point is that the PU coatings are irritant and toxic, and thus the PU coatings are not friendly to environment. The potential impact of the PU coatings on the environment should be carefully assessed before putting the PU coatings into use.

6.2.4 Alloy Coating

Many anti-corrosion organic coatings cannot be deployed under the supercritical CO_2 ($ScCO_2$) condition because the $ScCO_2$ is an extraordinary solvent that could dissolve a wide range of organic substances. Corrosion-resistant alloy (CRA) is a preferred choice to mitigate corrosion induced by $ScCO_2$, but the investment to manufacture CRA pipelines is too high. As an alternative to CRA pipelines, the alloy coating can effectively reduce the cost of CRA pipelines and also maintain an acceptable level of corrosion resistance. Nanocomposite alloy coatings have received a lot of interest as a result of the advancement of nanotechnology. A nanocomposite alloy coating is a composite substance made of insoluble dispersed-phase particles embedded in one or more matrix metals at the nanoscale. Due to their outstanding wear resistance, high hardness, high temperature oxidation resistance, and high corrosion resistance, nanocomposite alloy coatings can be applied in a wide range of scenarios [8].

Alloy coating under $ScCO_2$ environment is a new area, and research in this field has just begun. Sun et al. has investigated the interfacial processes and the corrosion performance of bilayer Ni–SiC coating, and they have found out that the corrosion inhibition efficiency of the Ni–SiC coating is higher than 80%, indicating a superior stability to prevent the corrosion [9]. Tang [10] has investigated the electrodeposited Ni–Co alloy coatings in sweet corrosion environments, and the results show that the fibre-like Ni–Co micro-structured coating exhibit 2 orders of magnitude greater CO_2 corrosion resistance than cone-like coating, and such coating could enhance the CO_2 corrosion resistance of X65 steel by 1000 times [10]. Ganesan [11] has posted a new post-supercritical CO_2 method to produce Ni–Cu alloy with high corrosion inhibition performance and long-term stability. Cui [8] has investigated the CO_2 corrosion resistance of the Ni–W–Y_2O_3–ZrO_2 nanocomposite coating, and they have found that the optimal coating structure resistant to CO_2 corrosion is Ni–W alloy as the matrix metal, and Y_2O_3 and ZrO_2 nanoparticles are added with concentrations of 10 g/L. Besides, there are studies on electroless nickel-based alloy coating, iron-based coating, Titanium alloy coating, etc.

6.2.5 Graphene Coating

Graphene anti-corrosion coatings have been a popular method of shielding metals against corrosion in recent years. The amount and size of coatings needed to protect metals can be reduced thanks to graphene coatings. A graphene coating has a stronger anti-corrosion impact than conventional anti-corrosion coatings while maintaining the metals' inherent thermal and electrical conductivity [12]. The thinnest two-dimensional (2D) carbon substance is graphene. Since Geim and Novoselov used a micro-mechanical stripping technique to create single-layer graphene for the first time, it has garnered a lot of interest from both the scientific and industrial communities [13]. Graphene is a structural monomer of carbon materials such as graphite, carbon nanotubes (CNTs), and fullerenes [14]. Numerous graphene-related applications, such as optical components [15], fuel cells [16], biological devices [17], and metal anti-corrosion coating, have been developed as a result of ongoing research on the performance of graphene. These applications profit from graphene's exceptional qualities, such as its great corrosion resistance, large specific surface area, high thermal conductivity, and good mechanical strength [12].

In 2011, it was discovered that pure graphene coverings have good corrosion resistance. According to Chen et al. [18], a CVD covering made of pure graphene shielded copper and a copper/nickel alloy against air oxidation. Chen et al. [18] developed a number of samples, including bare Cu and bare Cu/Ni alloy, as well as a Cu foil with a pure graphene anti-corrosion coating and a pure graphene anti-corrosion coating on it. To simulate an oxidative environment, these samples were heated for 4 h at 200 °C in the air and then submerged for 45 min in 30% H_2O_2. It was discovered that the Cu foil and the Cu/Ni alloy foil covered in the pure graphene covering did not have the same apparent color. For the Cu foil and the Cu/Ni alloy foil coated with pure graphene, only a little amount of white Cu_2O corrosion product was formed on the coating flaws while other regions of the coating exhibited no corrosion. In addition to protecting pure copper or pure nickel metals from corrosion, the pure graphene coating produced by CVD also offers corrosion protection for substrates such alloys, carbon steel, and stainless steel [19–21]. Zhang [21] provided evidence of the migration of graphene that was generated directly on the Cu foil to the Ni–Ti alloy's surface. In electrochemical testing, the pure graphene-coated Ni–Ti alloy showed greater corrosion potential and lower corrosion current than the uncoated Ni–Ti alloy. These findings demonstrated that the Ni–Ti alloy's corrosion resistance might be improved by a pure graphene layer. Additionally, pure graphene-coated alloys had a decreased rate of Ni ion release. Ye et al. formed a pure graphene coating on the surface of the carbon steel using laser induction and confirmed its corrosion resistance. With a corrosion rate of 0.05 mm/year, carbon steel coated with the pure graphene anti-corrosion coating had corrosion resistance that was comparable to stainless steel's (0.09 mm/year) [19].

6.2.6 Smart Coating

Smart anticorrosive coating works by spontaneously altering its material properties in response to an environmental stimulus, such as a corrosion agent [22]. Smart anticorrosive coating can recover or even enhance the anticorrosion ability of the coating with little manual intervention. This is due to the fact that the smart anti-corrosive coating gets immediately triggered off when the corrosion agents are present in the system, and the corrosion inhibitor/healing agents carried in the matrix of the anticorrosive coating actively fill the defect [1]. According to whether the coating is implanted with external repair agent, the smart coating can be divided into extrinsic and intrinsic self-healing smart anticorrosive coating [23]. For extrinsic self-healing smart anticorrosive coating, corrosion inhibitors such as cerium ion and praseodymium ion are added to the matrix of the smart anticorrosive coating [24]. The corrosion inhibitors can also be carried by nanocapsules. Healing agents and corrosion inhibitors contained in the nanocapsules may be released in response to a coating damage to assist in coating repair. The polymer matrix of the coating comprises self-healing structure for the intrinsic self-healing smart anticorrosive coating. Without implanting the external repair agent, the coating's molecular chain can be reassembled to repair coating damage with the use of external energy (heat, light, mechanical loading). The coating returns to its original structure and function after being stimulated externally [25]. In conclusion, smart anticorrosive coating not only has the function of active defense, but also has excellent passive barrier function. As a result, the coating's overall performance and service life are greatly increased [1].

6.2.7 Ion Implantation

Ion implantation is a process to inject impurity ions into the surfaces of materials using an ion accelerator [26]. Ion implantation is an emerging interdisciplinary technology related to atomic physics, nuclear physics and solid state physics. Ion implantation requires the conversion of an elemental atom into an ion, makes it accelerate under an electric field of tens to hundreds of kV per centimeter, and shoots the ion into the material surface in a vacuum target chamber after the ion obtains a very high speed. Entry of the ion into the material surface changes the surface composition, structure and properties of the material. Ion implantation technology grants researchers the freedom to choose the type of injected elements according to the specific needs and accurately control the energy and concentration of injected ions. Therefore, ion implantation is an ideal method for surface modification and has broad areas of applications [27].

Surface treatments of steel by ion implantation are generally directed to solve the problems of wear, friction, hardness, fatigue, corrosion and oxidation [26]. The use of ion implantation in this context can be dated back to the pioneering work originating from the group of Dearnaley and Hartley [28]. Recently, ion implantation has drawn great attention in the field of steel corrosion inhibition. The ion implantation approach is different from traditional coating approach to protect steel from corrosion, because traditional coating is applied outside the steel, while the ion implantation injects ions inside the steel. Therefore, the injected ions can be well bonded with the steel. Since corrosion always occurs at the material surface, strengthening the surface properties of steel can effectively improve the corrosion resistance of the steel. Ion implantation has the advantages of firm bonding with the steel matrix, formation of a super thin injection layer, and no ion solubility limitation. Therefore, ion implantation is able to achieve the purpose of steel corrosion resistance with little use of injected ions, which saves precious corrosion-resisting metal ions.

Ion implantation for corrosion inhibition has now been widely studied in the laboratories. For example, ion implantation has been applied to improve the corrosion resistance of iron, stainless steels, aluminum and its alloys [29], Mg alloys [30], lead, nickel [27], titanium and titanium alloys [31], biomaterials [32], etc. However, large scale on-site applications of ion implantation for steel corrosion inhibition are still rare.

6.3 Synthesis Approaches of Coatings

6.3.1 FBE Coating

The original material of FBE coating is the dry powder which remains unreacted under normal storage conditions. The essential components of the powder are resin, hardener, curing agent, fillers, extenders, and sometimes color pigment. The hardener and resin are combined to serve as the binder. The curing agent triggers the reaction, while pigments, fillers and extenders enable the FBE coating to have the desired properties. At usual temperatures for coating, ranging from 180 to 250 °C, the powder contents melt and turn to a liquid state. The liquid form of FBE flows onto the metallic surface and soon transform to the solid form through element cross-linking facilitated by the high temperature [33]. The process is named as "fusion bonding". It is important to note that there is no way to reverse the cross-linking that takes place during FBE production. The coating cannot return to its former state after curing has taken place. The FBE coating is referred to as a thermoset coating because it does not melt when exposed to further heat. The following procedures are used to create the FBE coating on the pipe's surface during the fusion bond epoxy process. The pipe's surface is first cleaned and heated; next, epoxy powder is electrostatically applied; last, FBE powder is melted onto the pipe's surface and dries quickly in a regulated and uniform thickness.

Nowadays, researchers are focusing on the enhancement of the corrosion resistance by adding nanomaterials as blocking fillers [34]. These nanomaterials include nanoparticles (silica, zinc, titanium, iron oxide, etc.), nanotubes (polymer/carbon) and nanofibers (carbon) which are considered as filling materials in epoxy coatings. The key problem of the application of nanomaterials is the dispersion in the polymer matrix. For example, the carbon nanotubes (CNTs) are difficult to be dispersed due to the strong interlayer van der Waals forces between the CNTs. In this regard, a series of approaches are developed to obtain a uniform dispersion quality of the CNTs in the polymer matrix including covalent and/or noncovalent functionalization methods, hybridization methods, and in-situ polymerization methods [34].

6.3.2 Alloy Coating

A typical strategy to prepare the alloy coating is electrodeposition due to the low cost and scalable range. As a well-known method, electrodeposition involves passing an electric current through conductive material that has been submerged in a solution.

The solution contains a salt of the metal that needs to be deposited on the surface of the conductive material to form in-situ metallic coatings. Additionally, electrochemical synthesis/deposition can be utilized to create thin layers of oxides and/or hydroxides on metals by carefully manipulating the synthesis conditions. The composition, morphology and texture of the film coating can be controlled by tuning the experimental parameters such as the potential, current density, deposition time, and plating solution composition [35].

Based on the type of the alloy and the technology advancement, the synthesis approaches of the coatings could be versatile. For example, Tang [36] has proposed an electrodeposition method of Ni–Co alloy coating on X65 steel with an area of 1 cm^2. The X65 steel was mechanically grounded down with abrasive SiC papers, followed by sequential polishing with 1 and 0.25 μm diamond pastes and ultrasonic cleaning in ethanol. After washing in a mixture of Na_2CO_3 (30 g/L) and NaOH (10 g/L) for 60 s at room temperature (20 °C), the X65 steel was activated in a 1:1 (v/v) HCl:H_2O solution for 30 s, rinsed in deionized water, and then put right into the electrodeposition bath. The electrodeposition bath comprised pH buffer (31 g/L H_3BO_3), CoC_2, and 238 g/L $NiC_2{\cdot}6H_2O$. NaOH solution was used to bring the bath's pH to 4.0. The creation of cone-shaped Ni-Co coatings required the use of the crystallization enhancer ethylenediamine (EDA, 200 g/L). To create Ni–Co coating with roughly 70% cobalt, the CoC_2 concentration was adjusted. By following those procedures, a Ni-Co alloy coating on X65 steel was produced. Sui [37] proposed an electrodeposit technique to deposit nanocrystalline Ni-based alloy coating on N80 steel. This technique included five processes: degreasing, pickling, activation, plating, and heat treatment (200–400 °C for 2–5 h). An iridium coating titanium electrode was used as the anode. N80 steel, which served as the cathode, was cut to a size of 50 mm × 10 mm × 3 mm. The N80 steel surface was abraded with SiC paper of decreasing roughness (up to 2000 grit) prior to the electrodeposition procedure. The Ni-bearing components in the plating solution were nickel sulphate ($c = 130$–230 g/L) and nickel carbonate ($c = 35$–55 g/L). The current density applied was 30–100 mA/cm^2, the deposition time was 75 min, the deposition temperautre was 50 ~ 80 °C, and the pH of the solution was maintained at 3.4–5.4.

6.3.3 Graphene Coating

Graphene coating has recently drawn a broad attention due to its superior anti-corrosion performance, but a suitable choice of a graphene making approach to produce an optimal yield must first be carefully evaluated in order to optimize the wide variety of graphene applications in diverse fields. The four major techniques for making graphene are micromechanical stripping, silicon carbide (SiC) epitaxial growth, redox reaction, and CVD reaction [12]. Novoselov [13] used the micromechanical stripping technique, or the act of stripping graphite with a tape, to produce single-layer graphene for the first time. Tape was used to directly peel off a layer of graphite, which was then repeatedly deposited on the tape. The graphite layer consequently got thinner and thinner. Finally, the tape was attached to the substrate, and a single layer of graphene was transferred to the surface of the substrate. This straightforward graphene coating preparation procedure yielded graphene of excellent quality. However, the graphene coating could not be synthesized in vast amounts by micromechanical stripping, and the coating was only achieved in tiny areas. A redox method has been created to get over the drawbacks of the micromechanical stripping technique. This approach involves oxidizing graphite with potassium permanganate, concentrated sulfuric acid, and concentrated nitric acid to produce graphene oxide (GO), which is then separated from the graphite layers by oxygen-bearing functional groups. Ultrasonic-assisted dispersion is used to create the GO sheets that are distributed in solution. Strong reducing chemicals such sodium borohydride and hydrazine hydrate are then used to deoxygenate the GO to produce graphene [12]. Large-scale graphene preparation has been carried out using this easy, affordable technique, and it has become popular. However, during the strong oxidation and reduction processes, the structure of graphene may be damaged, resulting in defects and poor quality of graphene [12]. In short, industry-scale anti-corrosion applications of graphene coating are still limited due to a lack of efficient approaches to produce the required yield with low cost.

References

1. Cui, G., Bi, Z., Wang, S., Liu, J., Xing, X., Li, Z., Wang, B.: A comprehensive review on smart anti-corrosive coatings. Prog. Org. Coat. **148**, 105821 (2020)
2. Qian, Y., Li, Y., Jungwirth, S., Seely, N., Fang, Y., Shi, X.: The application of anti-corrosion coating for preserving the value of equipment asset in chloride-laden environments: a review. Int. J. Electrochem. Sci **10**, 10756–10780 (2015)
3. Verma, C., Olasunkanmi, L.O., Akpan, E.D., Quraishi, M.A., Dagdag, O., El Gouri, M., Sherif, E.-S.M., Ebenso, E.E.: Epoxy resins as anticorrosive polymeric materials: a review. React. Funct. Polym. **156**, 104741 (2020)
4. Bhadouria, A.S., Kumar, A., Raj, D., Verma, A., Singh, S., Tripathi, P., Kumar, Y., Sinha, A.S.K., Tripathi, N.M., Yeneneh, A.M., Dwivedi, D.: Corrosion mitigation in oil reservoirs during CO_2 injection using nanomaterials, in Nanotechnology for CO_2 utilization in oilfield applications, pp. 127–146, Chap. 8. Gulf Professional Publishing, Amsterdam (2022)
5. Samimi, A., Zarinabadi, S.: An analysis of polyethylene coating corrosion in oil and gas pipelines. In: First International Conference of Oil, Gas, Petrochemical and Power Plant, USA (2011)
6. Wasim, M., Djukic, M.B.: External corrosion of oil and gas pipelines: a review of failure mechanisms and predictive preventions. J. Nat. Gas Sci. Eng. **100**, 104467 (2022)
7. Marathe, R., Tatiya, P., Chaudhari, A., Lee, J., Mahulikar, P., Sohn, D., Gite, V.: Neem acetylated polyester polyol—renewable source based smart PU coatings containing quinoline (corrosion inhibitor) encapsulated polyurea microcapsules for enhance anticorrosive property. Ind. Crops Prod. **77**, 239–250 (2015)
8. Cui, G., Bi, Z., Liu, J., Wang, S., Li, Z.: New method for CO_2 corrosion resistance Ni-W-Y_2O_3-ZrO_2 nanocomposite coatings. Ceram. Int. **45**(5), 6163–6174 (2019)
9. Kim, S.K., Yoo, H.J.: Formation of bilayer Ni–SiC composite coatings by electrodeposition. Surf. Coat. Technol. **108–109**, 564–569 (1998)
10. Tang, H., Zhang, D., Li, Y.: Research on self-lubricating coating with electroless composite (Ni-Co-P)-MoS~2 coating. Electroplat. Pollut. Control **26**(4), 23 (2006)

11. Ganesan, M., Liu, C.-C., Pandiyarajan, S., Lee, C.-T., Chuang, H.-C.: Post-supercritical CO_2 electrodeposition approach for Ni-Cu alloy fabrication: an innovative eco-friendly strategy for high-performance corrosion resistance with durability. Appl. Surf. Sci. **577**, 151955 (2022)
12. Cui, G., Bi, Z., Zhang, R., Liu, J., Yu, X., Li, Z.: A comprehensive review on graphene-based anti-corrosive coatings. Chem. Eng. J. **373**, 104–121 (2019)
13. Novoselov, K.S., Geim, A.K., Morozov, S.V., Jiang, D.-E., Zhang, Y., Dubonos, S.V., Grigorieva, I.V., Firsov, A.A.: Electric field effect in atomically thin carbon films. Science **306**(5696), 666–669 (2004)
14. Geim, A.K., Novoselov, K.S.: The rise of graphene. Nat. Mater. **6**(3), 183–191 (2007)
15. Kim, J.T., Choi, H., Shin, E., Park, S., Kim, I.G.: Graphene-based optical waveguide tactile sensor for dynamic response. Sci. Rep. **8**(1), 1–6 (2018)
16. Akbar, K., Kim, J.H., Lee, Z., Kim, M., Yi, Y., Chun, S.-H.: Superaerophobic graphene nano-hills for direct hydrazine fuel cells. NPG Asia Mater. **9**(5), e378 (2017)
17. Goldsmith, B.R., Locascio, L., Gao, Y., Lerner, M., Walker, A., Lerner, J., Kyaw, J., Shue, A., Afsahi, S., Pan, D.: Digital biosensing by foundry-fabricated graphene sensors. Sci. Rep. **9**(1), 1–10 (2019)
18. Chen, X.M., Song, S.H., Weng, L.Q., Liu, S.J.: Solute grain boundary segregation during high temperature plastic deformation in a Cr–Mo low alloy steel. Mater. Sci. Eng. A **528**, 7663–7668 (2011)
19. Ye, X., Lin, Z., Zhang, H., Zhu, H., Liu, Z., Zhong, M.: Protecting carbon steel from corrosion by laser in situ grown graphene films. Carbon **94**, 326–334 (2015)
20. Zhu, M., Du, Z., Yin, Z., Zhou, W., Liu, Z., Tsang, S.H., Teo, E.H.T.: Low-temperature in situ growth of graphene on metallic substrates and its application in anticorrosion. ACS Appl. Mater. Interfaces **8**(1), 502–510 (2016)
21. Zhang, L., Duan, Y., Gao, Z., Ma, J., Liu, R., Liu, S., Tu, Z., Liu, Y., Bai, C., Cui, L.: Graphene enhanced anti-corrosion and biocompatibility of NiTi alloy. NanoImpact **7**, 7–14 (2017)
22. Poornima Vijayan, P., Al-Maadeed, M.: Self-repairing composites for corrosion protection: a review on recent strategies and evaluation methods. Materials, **12**(17), 2754 (2019)
23. Wang, X., Guo, Y., Su, J., Zhang, X., Wang, Y., Tan, Y.: Fabrication and characterization of novel electrothermal self-healing microcapsules with graphene/polymer hybrid shells for bitumenious material. Nanomaterials **8**(6), 419 (2018)
24. Lakshmi, R.V., Yoganandan, G., Kavya, K.T., Basu, B.J.: Effective corrosion inhibition performance of Ce^{3+} doped sol–gel nanocomposite coating on aluminum alloy. Prog. Org. Coat. **76**(2), 367–374 (2013)
25. Garcia, S.J.: Effect of polymer architecture on the intrinsic self-healing character of polymers. Eur. Polym. J. **53**, 118–125 (2014)
26. Townsend, P.: Ion implantation—an introduction. Contemp. Phys. **27**(3), 241–256 (1986)
27. Liao, S.: Application of Ion Implantation in Corrosion and Protection of Lead, Nickel and Aluminium. Chongqing University, Chongqing (2002)
28. Hartley, N., Swindlehurst, W., Dearnaley, G., Turner, J.: Friction changes in ion-implanted steel. J. Mater. Sci. **8**, 900–904 (1973)
29. McCafferty, E.: Effect of ion implantation on the corrosion behavior of iron, stainless steels, and aluminum—a review. Corrosion **57**(12), 1011–1029 (2001)
30. Chen, C., Wang, M., Liu, Y.: New surface modification technology for Mg alloys. Corros. Sci. Prot. Technol. **16**(4), 215–217 (2004)
31. Rautray, T.R., Narayanan, R., Kwon, T.Y., Kim, K.H.: Surface modification of titanium and titanium alloys by ion implantation. J. Biomed. Mater. Res., Part B **93**(2), 581–591 (2010)
32. Lu, T., Qiao, Y., Liu, X.: Surface modification of biomaterials using plasma immersion ion implantation and deposition. Interface focus **2**(3), 325–336 (2012)
33. Malik, A.U., Andijani, I., Ahmed, S., Al-Muaili, F.: Corrosion and mechanical behavior of fusion bonded epoxy (FBE) in aqueous media. Desalination, **150**(3), 247–254 (2002)
34. Chen, S., Wang, X., Zhu, G., Lu, Z., Zhang, Y., Zhao, X., Hou, B.: Developing multi-wall carbon nanotubes/fusion-bonded epoxy powder nanocomposite coatings with superior anti-corrosion and mechanical properties. Colloids Surf. A **628**, 127309 (2021)

35. Alagu Sundara, P., Srinivasan, R. G., Palani, S., Selvam, M.: Surface modification on AZ31B Mg alloy for improved corrosion resistance and hardness by thermal spray aluminium coating. Materials Today: Proceedings 2586–2592 (2023)
36. Tang, R., Joshi, G.R., Zhao, H., Venkateswaran, S.P., Withers, P.J., Xiao, P.: The influence of electrodeposited Ni-Co alloy coating microstructure on CO_2 corrosion resistance on X65 steel. Corros. Sci. **167**, 108485 (2020)
37. Sui, Y., Sun, C., Sun, J., Pu, B., Ren, W., Zhao, W.: Stability of an electrodeposited nanocrystalline Ni-based alloy coating in oil and gas wells with the coexistence of H_2S and CO_2. Materials **10**(6), 632 (2017)

Chapter 7
Corrosion Control (III): Corrosion Inhibitors

Manguang Gan

7.1 General Introduction

During the process of capturing and transporting CO_2 by pipeline, the solubility of CO_2 in water varies considering the CO_2 partial pressure and temperature. Due to the inefficient removal of water from the source gas, the CO_2 is saturated with dispersed water droplets and the dissolution of CO_2 will lead to the formation of carbonic acid. The steel (including pipeline, tubing, casing) would not corrode in dry CO_2, but corrode in carbonic acid (carbonic acid is formed in the process of CO_2 capture [5–8], and storage [9]), and the corrosion rate depends on temperature, CO_2 partial pressure, solution composition, impurity, material characteristics, pH, solution flow rate, rates of the participating anodic and cathodic reactions, etc. [10–30]. There are three typical control techniques for CO_2 corrosion resistance in CCUS: selecting corrosion-resistant materials [31], inner wall coating or lining [32], and adding corrosion inhibitors [33]. Among those anti-corrosion methods, the use of corrosion inhibitors seems to be the optimal choice. Adding corrosion inhibitors is a relatively economical, effective, and versatile metal corrosion control method, which is suitable for the whole process of CCUS [28, 34]. Corrosion inhibitors are chemical substances or compounds that form a hydrophobic protective film on the surface of metal materials through physical and chemical actions that protect the metals from interacting with the corrosive medium. Any substance that can significantly reduce the corrosion rate by adding a small amount to the corrosive medium is called a corrosion inhibitor. Adding corrosion inhibitors can effectively mitigate the corrosion of downhole transportation and pipelines, if the dosage is properly adjusted

M. Gan (✉)
State Key Laboratory of Geomechanics and Geotechnical Engineering, Institute of Rock and Soil Mechanics, Chinese Academy of Sciences, Wuhan 430071, China
e-mail: cumtgmg@163.com

University of Chinese Academy of Sciences, Beijing 100049, China

L. Zhang (ed.), *Corrosion in CO_2 Capture, Transportation, Geological Utilization and Storage*, Engineering Materials, https://doi.org/10.1007/978-981-99-2392-2_7

according to the changes in oil, water, and gas components during the production of oil and gas and the process of geological CO_2 storage.

According to the related literature[9], the corrosion rate of carbon steel under high CO_2 pressure without a $FeCO_3$ protective film was found to be considerably high (20 mm/year), while the corrosion rate can be reduced to a lower value (0.2 mm/year) due to the fast formation of a $FeCO_3$ protective film after using a corrosion inhibitor. In addition, based on the field data (pressure: 5.92 ~ 12.2 MPa; temperature: 40 ~ 44 °C) [9], corrosion rates were about 6.35 mm/year on the N-80 tubing and 0.13 mm/year on J-55 tubing after 10 months of production. Therefore, some oilfields use corrosion-resistant materials, such as stainless steel or non-ferrous alloy oil/casing materials to mitigate the problem of metal corrosion, but the high cost severely limits the possibility of providing corrosion protection in terms of material selection. In addition, the treatment process of the coating or lining of the inner wall is very complicated, and the metal is easily damaged or defective, which would lead to severe local corrosion. Therefore, this anti-corrosion process is often used in conjunction with the addition of corrosion inhibitors to ensure the effectiveness of the anti-corrosion technology.

The inhibition mechanisms of corrosion inhibitors mainly include the adsorption mechanism and the film formation mechanism. The adsorption mechanism points out that most organic corrosion inhibitors are surfactants, which are composed of hydrophilic and oleophobic polar groups and lipophilic and hydrophobic non-polar groups. On the metal surface, the corrosion-causing agents such as water molecules and hydrogen ions are excluded by adsorbed organic corrosion inhibitors to prevent corrosion. The film-forming mechanism points out that the ions in the medium or metals react with the corrosion inhibitor to form a film that can prevent the corrosion process and play a role in corrosion inhibition. This film is divided into two types, namely, passivation film and precipitation film.

Based on the inhibition mechanism of the corrosion inhibitor, it is known that the corrosion inhibitor usually possesses one or several functional protective groups, which enable the corrosion inhibitor to adsorb on the metal surface and undergo a series of physical and chemical changes. The polar group has strong adsorption, and different corrosion inhibitor groups have different impacts on the effectiveness of metal protection. For example, the hydrocarbon group is the non-polar part of the corrosion inhibitor, its length and structure have an impact on the rust resistance of the corrosion inhibitor. On the one hand, the more carbon atoms in the hydrocarbon group, the larger the molecular volume, the thicker and tighter the adsorption layer, and the better the anti-corrosion effect. On the other hand, the longer the hydrocarbon group, the lower the water solubility, the larger the molecular motion resistance, the slower the adsorption speed, and the longer the induction period, which affects the corrosion inhibition effect. In short, the factors that impact the inhibition performance of the corrosion inhibitor include the environment where the corrosion inhibitor is located, the essential characteristics of the metal surface, the structure and properties of the corrosion inhibitor itself (including the functional groups and its molecule size), the method of adsorption, the form of bonding with the metal, the area covered on the metal surface, and the interfacial potential between the metal and the adsorption layer.

7.1.1 The Process of Steel Corrosion and Characterization of Corrosion Degree

In the wellbore system of a typical CCUS site, as shown in Fig. 7.1, contact between the CO_2-dissolved fluid and the steel casing can occur at the micro-annulis along the cement-casing interface, stress cracks within the cement, and uncemented regions within the annulus space. When CO_2 encounters water and forms carbonic acid on the steel surface, the corrosion reactions will occur on the surface of the steel, which will be discussed below.

The equilibria of an aqueous CO_2 system and the electrochemical corrosion process of steel are shown in Fig. 7.2. The stepwise process of CO_2 corrosion involves: (i) dissolution of CO_2, (ii) CO_2 hydration and formation of carbonic acid, and (iii) carbonic acid dissociation. These processes include [6, 36]:

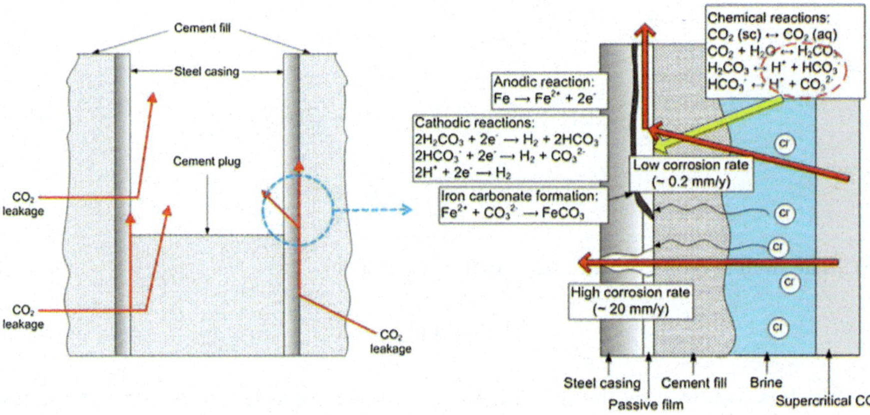

Fig. 7.1 Loss of wellbore integrity due to CO_2 corrosion [9]

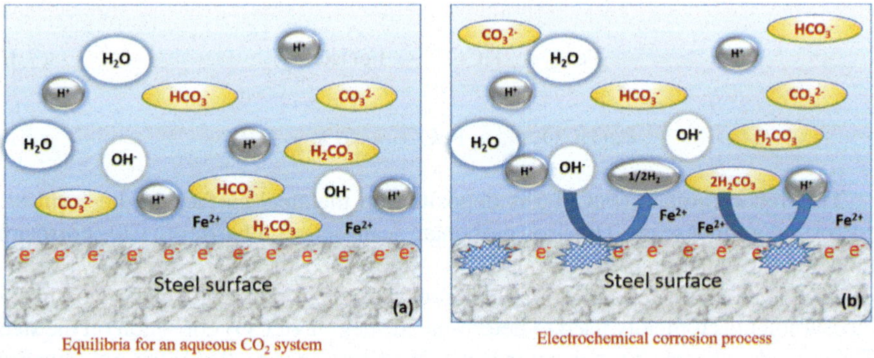

Fig. 7.2 Schematic diagram of an aqueous CO_2 system and the process of CO_2 corrosion [35]

(1) The formation of bicarbonate and carbonate ions in two steps after saturation of condensate with CO_2.

$$CO_{2(g)} \leftrightarrow CO_{2(aq)} \tag{7.1}$$

$$CO_{2(g)} + H_2O_{(l)} \leftrightarrow H_2CO_{3(aq)} \tag{7.2}$$

$$H_2CO_3 \leftrightarrow H^+ + HCO_3^- \tag{7.3}$$

$$HCO_3^- \leftrightarrow H^+ + CO_3^{2-} \tag{7.4}$$

(2) In the next stage, the cathodic reaction may occur either by direct reduction of hydrogen ions, or the reduction of carbonic or carbonate ions:

$$2H^+ + 2e^- \leftrightarrow H_2 \tag{7.5}$$

$$2H_2CO_3 + 2e^- \rightarrow H_2 + 2HCO_3^- \tag{7.6}$$

$$2HCO_3^- + 2e^- \rightarrow H_2 + 2CO_3^{2-} \tag{7.7}$$

(3) The final stage is the anodic dissolution of iron:

$$Fe \rightarrow Fe^{2+} + 2e^- \tag{7.8}$$

Which is followed by the precipitation via a one-stage reaction with carbonates, or a two-stage reaction with bicarbonates:

$$Fe^{2+} + CO_3^{2-} \rightarrow FeCO_3 \tag{7.9}$$

$$Fe^{2+} + 2HCO_3^- \rightarrow Fe(HCO_3)_2 \tag{7.10}$$

$$Fe(HCO_3)_2 \rightarrow FeCO_3 + CO_2 + H_2O \tag{7.11}$$

The iron carbonate ($FeCO_3$) is deposited on the steel surface in the form of a corrosion product scale. This iron carbonate scale acts as a barrier to CO_2 corrosion, reducing the general corrosion rate [37].

As for the characterization of steel corrosion degree in CCUS, the weight-loss measurement is typically used to determine the degree of CO_2 corrosion on the steel. The mass loss due to CO_2 corrosion is usually determined by the difference in weight before exposure and after cleaning. The corrosion rate is determined by the Eq. (7.12) [36]:

$$CR = \frac{8.76 \times 10^4 \times \Delta m}{\rho \times A \times T} \quad (7.12)$$

where CR is the corrosion rate of steel sample in mm/year, Δm is the weight loss in grams, ρ is the density of the steel sample in g/cm^3, A is the exposed surface area in cm^2 and T is the immersion time in hours.

The efficiency of corrosion inhibitor is determined by Eq. (7.13):

$$IE\ (\%) = \frac{CR_0 - CR_{inh}}{CR_0} \times 100 \quad (7.13)$$

where IE (%) is inhibition efficiency, CR_{inh} and CR_0 are the corrosion rates of steel with and without CO_2 corrosion inhibitor in mm/year.

7.1.2 Classification of Corrosion Inhibitors

The corrosion process of metals in electrolyte solution consists of two conjugated electrochemical reactions, namely, the anodic reaction and the cathodic reaction. The corrosion rate of the metal can be reduced if the corrosion inhibitor inhibits either or both anodic and cathodic reactions. Therefore, CO_2 corrosion inhibitors can be divided into anodic corrosion inhibitors, cathodic corrosion inhibitors, and mixed corrosion inhibitors according to the influence of a given CO_2 corrosion inhibitor on the anodic or cathodic reaction speed of the corrosion process. In addition, according to the situation that the corrosion inhibitor changes the surface state of the corroded metal, it can be divided into film-forming corrosion inhibitor and adsorption corrosion inhibitor [38].

The following discusses the classification by the effect of corrosion inhibitors on the anodic or cathodic reaction of the corrosion process.

Anodic corrosion inhibitor: Anode corrosion inhibitor can increase anodic polarization in the process of metal corrosion and move the corrosion potential to the positive direction. Generally, the anions of the anodic corrosion inhibitor move to the surface of the metal anode, covering a layer of oxide or hydroxide film on the metal surface, thereby hindering the migration of metal ions to the solution, and slowing down the corrosion rate by inhibiting the anode reaction process. Anodic corrosion inhibitors mainly include nitrates, nitrites, phosphates, silicates, etc., and generally serve as corrosion inhibition in the form of a passivation protective film. When using an anodic corrosion inhibitor, the dosage must be enough. If the dosage is insufficient, the corrosion inhibitor cannot fully cover the surface of the anode, so that the area of the anode exposed to the medium is much smaller than the area of the cathode, forming a corrosion battery with a small anode and a large cathode, which intensifies the pitting corrosion of metal. Therefore, anode corrosion inhibitors are also known as hazardous corrosion inhibitors.

Cathodic corrosion inhibitor: The cathodic corrosion inhibitor has a significantly greater impact on the cathode than the anode in the process of corrosion. Cathodic corrosion inhibitor plays a role in the cathode of the corrosion system, and the product forms a film in the cathode area, and the thickness of the film increases with the progress of the reaction, thereby hindering the diffusion of corrosion agents at the interface of the electric double layer. Typically, cathodic corrosion inhibitors are cations that move to the cathode surface to form a chemical or electrochemical precipitation film, thereby inhibiting metal corrosion. Even if the amount of this type of corrosion inhibitor is insufficient, it will not accelerate corrosion, so it is called a safe corrosion inhibitor. The most widely used cathodic corrosion inhibitors are zinc carbonate, phosphate and hydroxide, calcium carbonate, and phosphate.

Mixed corrosion inhibitor: The organic compounds containing N, S, –OH and surface activity are called mixed corrosion inhibitors. There are groups with different properties in the organic structure of such corrosion inhibitors, and these groups are chemically bonded to the metal surface. Monomolecular films are formed and adsorbed on metal surfaces. The mixed corrosion inhibitor can inhibit the cathodic process and the anodic process at the same time and form a film in both the anodic and cathodic regions, thereby preventing the solution and the dissolved oxygen in the solution from diffusing to the metal surface. After adding the mixed corrosion inhibitor, although the corrosion potential did not change significantly, the corrosion current decreased significantly. Alkaloids, cyclic imines, etc. are typically mixed corrosion inhibitors.

Besides, the classification of the surface state of the corroded metal by the corrosion inhibitor is also discuss below.

Film-forming corrosion inhibitors: Film-forming corrosion inhibitors are mainly inorganic substances (such as chromates, nitrites, etc.), and are mainly used in medium-sized media. These corrosion inhibitors have good performance, but these inhibitors are often used in large amounts, and the feasibility is poor. When the amount of corrosion inhibitor is insufficient, severe local corrosion may occur.

Adsorption-type corrosion inhibitors: At present, most of the corrosion inhibitors are adsorption-type corrosion inhibitors, such as chain organic amines and their derivatives, imidazolines and their salts, imidazoline derivatives quaternary amine salts, rosin amine derivatives, other organic compounds, etc. Among them, imidazoline corrosion inhibitors are the most widely used corrosion inhibitors and have obvious corrosion inhibition effects on systems containing CO_2 or H_2S. Ammonium salt and quaternary ammonium salt corrosion inhibitors are mainly relying on nitrogen atom adsorption, which is widely used in adsorption film formation. Organophosphorus corrosion inhibitor is a kind of efficient and widely used corrosion inhibitor, which is mainly used in industrial water, oil, natural gas extraction, and refining of metal anti-corrosion. Alkynols are important steel corrosion inhibitors under high temperature and concentrated acid conditions. The molecules of alkynols have both polar groups such as –OH and C≡C, and non-polar groups such as hydrocarbon groups. The π electrons on the alkyne bond are metallic, which makes it easy to adsorb on the metal surface, and the non-polar group is at the end away from the metal surface, which shields the corrosive medium.

7.2 Synthesis Approaches

Figure 7.3 shows the general structures of some corrosion inhibitors for CO_2 media, and the organic inhibitors efficiently coordinate with metal ions to form metal-inhibitor complexes.

The typical CO_2 corrosion inhibitor includes imidazoline, quaternary ammonium salt, amide, phosphorus-containing compound, alcohol corrosion inhibitors, and compound corrosion inhibitors with synergistic effects. The synthesis approaches of each corrosion inhibitor are discussed below:

Imidazoline: Imidazoline is generally obtained by the condensation reaction of organic acid and organic polyamine. Imidazoline corrosion inhibitor is a nitrogen-containing five-membered heterocyclic compound with no special irritating odor, good thermal stability and low toxicity, and has an obvious corrosion inhibition effect on systems containing CO_2 or H_2S. The imidazoline corrosion inhibitors that resist CO_2 corrosion in wells are adsorbed by nitrogen atoms and are mainly divided into oil-soluble and water-soluble types. This type of corrosion inhibitor generally consists of three parts: a five-membered heterocyclic ring containing nitrogen, a branched chain R_1 (such as amide functional group, amine functional group, hydroxyl group, etc.) on the heterocyclic ring, and a long hydrocarbon branched chain R_2

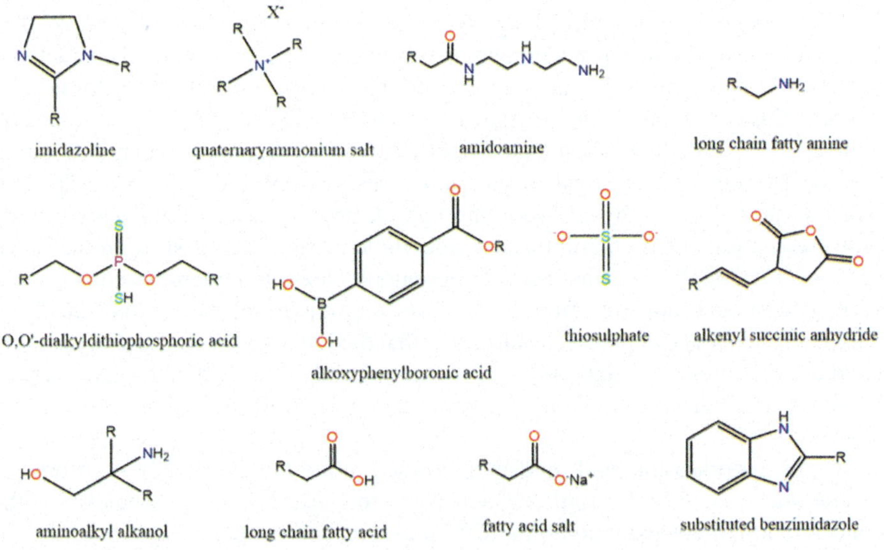

Fig. 7.3 General structure of corrosion inhibitors for CO_2 environments [35]

(generally an alkyl group). It is a corrosion inhibitor widely used in CCUS, oil and natural gas production, and has an obvious corrosion inhibition effect on systems containing CO_2. When the imidazoline corrosion inhibitor is compounded, it can produce a synergistic effect, so that the adsorption of different inhibitor molecules or ions can attract each other, increase the surface coverage, and thus improve the corrosion inhibition effect. Although imidazolines and their derivatives have been widely used as corrosion inhibitors and their effects are satisfactory, there is still much work to do for improvement in their corrosion inhibition efficiency and toxicity control.

Generally, the reaction formula of imidazoline is as follows:

(1) amidation reaction

$$\mathrm{RCOOH} + \mathrm{NH_2(CH_2CH_2NH)}_n\mathrm{CH_2CH_2NH_2} \xrightarrow{-\mathrm{H_2O}} \mathrm{RCONH(CH_2CH_2NH)}_n\mathrm{CH_2CH_2NH_2}$$

(2) cyclization reaction

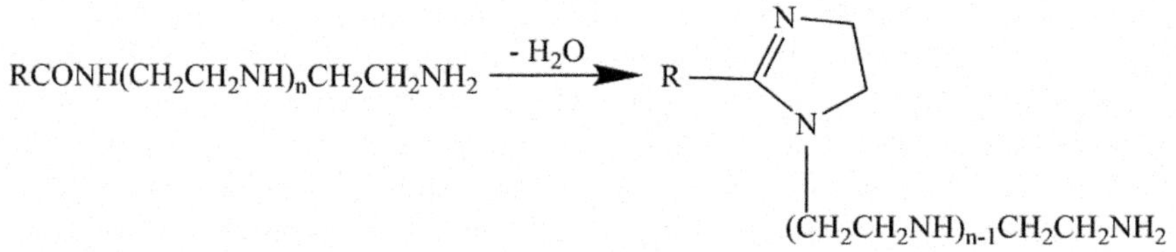

The function of corrosion inhibitors in the CO_2 atmosphere is illustrated in Fig. 7.4, which takes a typical long-chain im-based molecule as an example. The effectiveness of adsorption and protection is influenced by the structural characteristics of the inhibitor molecule, including functional groups, alkyl chains, heteroatoms, benzene rings, and -bond conjugation. Two types of corrosion processes are seen on a metallic substrate that is being attacked by corrosion: (i) anodic metal dissolution and (ii) cathodic hydrogen gas evolution. Corrosion inhibitors can be adsorbed on metal surfaces by: (i) physisorption through electrostatic interactions between the protonated atoms of the inhibitor and cathodic corrosion active sites on the metal surface; (ii) through lone pairs sharing heteroatoms electrons undergo chemisorption; (iii) electron back donation from metal surface atoms to inhibitor molecules [35]. Quantum chemical calculation results show that the imidazoline ring and heteroatom are the active sites of the inhibitor[39]. It can be firmly adsorbed on the iron surface by donating electrons to iron atoms and accepting electrons on the 3d orbits of iron atoms.

In short, imidazoline and imidazoline-based inhibitors are widely used in CO_2 conditions [24, 39–48], and the anti-corrosion effect mainly depends on the concentration of corrosion inhibitor, CO_2 pressure, and temperature.

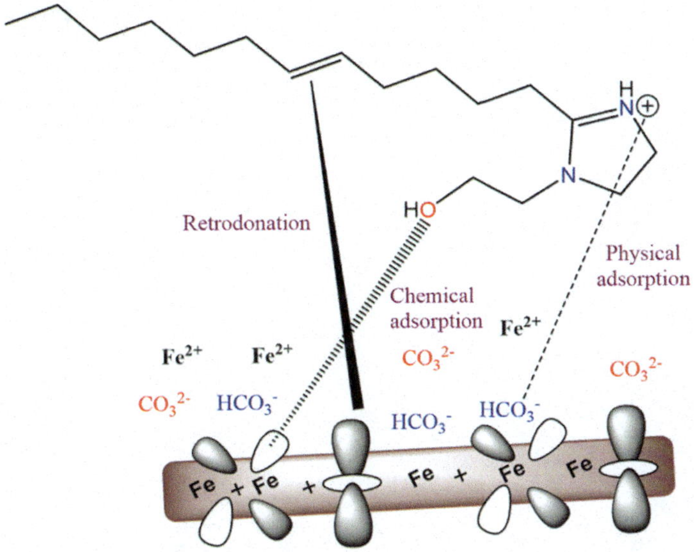

Fig. 7.4 General mechanism of adsorption and inhibition of imidazoline-based corrosion inhibitors on steel in CO_2 environment [35]

Quaternary ammonium salt corrosion inhibitor Quaternary ammonium salt is formed when the nitrogen atom in a nitrogen compound is converted into a pentavalent form. This corrosion inhibitor mainly relies on nitrogen atom adsorption and is widely used in wells for adsorption and film formation. It not only contains O, N, and large π chain active radicals, but also contains synergistic quaternary ammonium cation and halogen anion. It is a kind of adsorption film inhibitor with a low dosage and good effect, but may toxicity the environment [49]. The quaternary ammonium cation is adsorbed on the metal surface with a negative charge, forming a dense and complete coating film on the metal surface, which can effectively inhibit the anodic reaction [50]. In addition, the anion on the quaternary ammonium salt has a great influence on the electrostatic adsorption of the cationic corrosion inhibitor.

As a common quaternary ammonium salt corrosion inhibitor, the synthesis processes of quaternary ammonium salt corrosion inhibitor of [11-methoxy-11-oxo-N, N, N tripropylundecan-1-aminium bromide] and [11-((2-((2-aminoethyl) amino) ethyl) amino)-11-oxo-N, N, N tripropylundecan-1-aminium bromide] are shown in Fig. 7.5. The adsorption and inhibition behavior of corrosion inhibitor [11-((2-((2-aminoethyl) amino) ethyl) amino)-11-oxo-N, N, N tripropylundecan-1-aminium bromide] on steel surface is shown in Fig. 7.6. The synthesis process is complicated and the detail can be found in the related literature [50].

Besides, Zhao [51] investigated the effect of quinoline quaternary ammonium salt and Gemini surfactant 1,3-bis(dodecyl dimethyl ammonium chloride)-2-propanol for mild steel in H_2S and CO_2 saturated brine through polarization test, EIS and XPS.

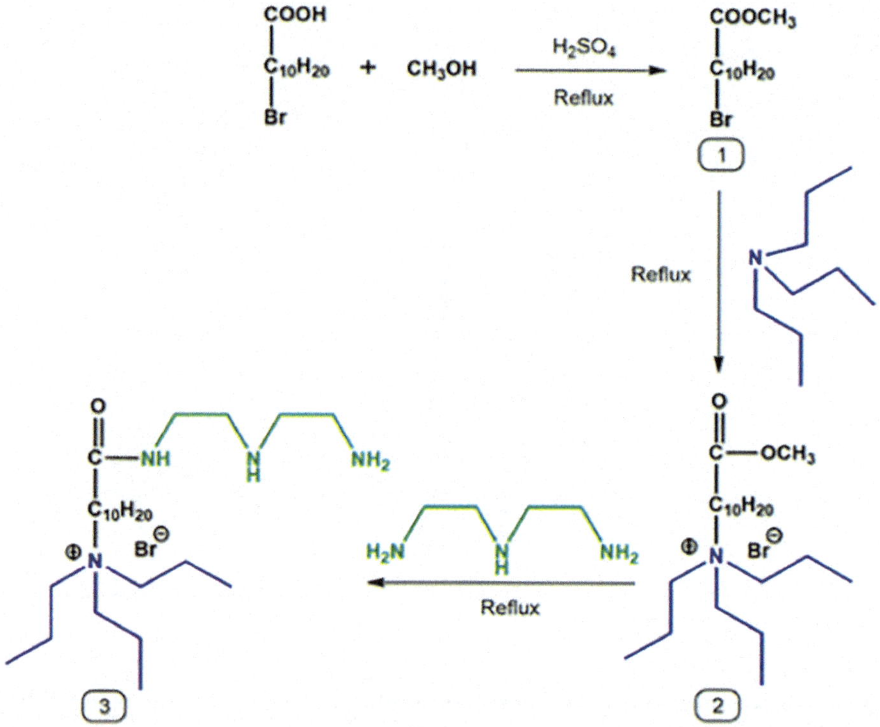

Fig. 7.5 Synthesis of corrosion inhibitors [11-methoxy-11-oxo-N, N, N tripropylundecan-1-aminium bromide] (2) and [11-((2-((2-aminoethyl) amino) ethyl) amino)-11-oxo-N, N,N tripropylundecan-1-aminium bromide] (3) [50]

The synergistic mechanism of the solution containing quinoline quaternary ammonium salt and Gemini depends on the concentration of Gemini and the competitive adsorption of these two compounds on the steel surface explains the synergistic mechanism. Khodyrev [52] tested the ammonium salts of O, O′-dialkyldithiophosphoric acids $(RO)_2$ P(S) S^- HN^+ $R'R''_2$ as inhibitors for CO_2 corrosion of mild steel and found that high inhibition efficiency has been found for all compounds (g = 70 ~ 99%) at very low concentrations (0.25–5 mg/l).

Amide corrosion inhibitor Amide corrosion inhibitor belongs to an organic amine corrosion inhibitor. Due to the existence of amide bonds in the molecules, it has good hydrolysis resistance, stability, low toxicity, and biodegradation in a wide range of pH values. It can be used in acidic media, neutral media, and atmospheric corrosion media, especially suitable for CO_2 corrosion resistance in the CCUS. There are many chemical routes for the preparation of amide corrosion inhibitors. The main route is the synthesis of fatty acid amide by reaction of fatty acid and ammonia. In the absence of a catalyst, the synthesis of amide by reaction of fatty acid and ammonia needs to be carried out under high temperature and high pressure.

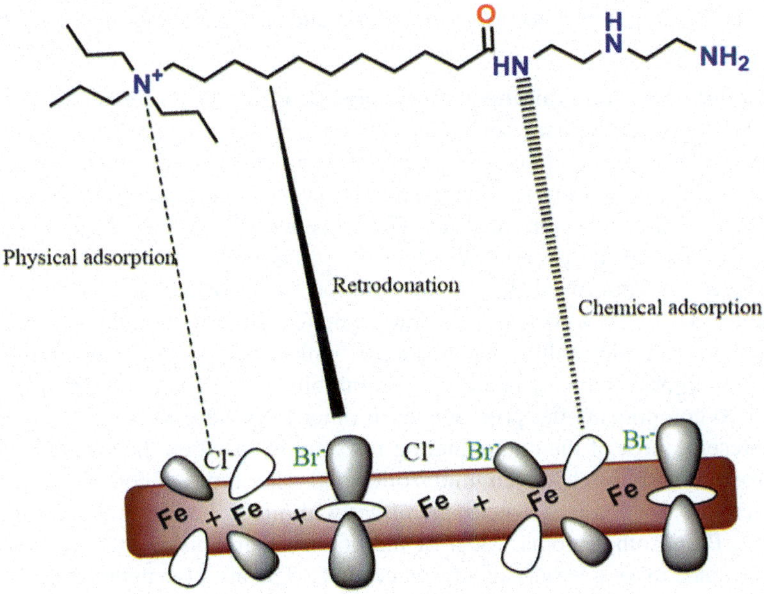

Fig. 7.6 Schematic of the mechanism of adsorption and inhibition behavior of corrosion inhibitor [11-((2-((2-aminoethyl) amino) ethyl) amino)-11-oxo-N, N, N tripropylundecan-1-aminium bromide] on steel surface [50]

The corrosion inhibitor of N-[2-[(2-aminoethyl) amino] ethyl]-9-octadecenamide on the mild steel was tested in CO_2-saturated 5% NaCl solution [53]. The inhibitors block surface sites at low concentrations and form a protective barrier against highly corrosive ions by modifying the adsorption mechanism.

Phosphorus-containing compound corrosion inhibitor Organophosphorus corrosion inhibitor is a kind of highly efficient and widely used corrosion inhibitor. However, because of the tendency of phosphorus compounds to cause water eutrophication, their industrial applications are increasingly limited. Many metal ions such as Ca, Mg, and Fe are often precipitated in the form of sulfate and carbonate in the water of wells, resulting in scaling and corrosion of pipelines and equipment, and thus affecting the normal operation and use of equipment. Organophosphate corrosion inhibitor is one of the most effective scale inhibitors at present, it can inhibit the formation of scale and the corrosion of metal matrix in water with high Ca^{2+}, CO_3^{2-} content, and high pH by adsorption of phosphorus atoms.

Alcohol corrosion inhibitors Alkynyl alcohol compounds are important steel corrosion inhibitors under high temperature and concentrated acid conditions, especially when mixed with nitrogen-containing compounds, they can be used in high-temperature environments above 100 °C. The molecule of alkynol has both polar groups and non-polar groups, and the π electron on its alkyne bond is metallic, which makes it easy to adsorb on the metal surface, and the non-polar group is at the end away from the metal surface. Adsorption and shielding can effectively prevent pitting corrosion.

Compound corrosion inhibitor with synergistic effect The compounding of corrosion inhibitors is a promising research direction in the development of new corrosion inhibitors. Due to the complexity of the metal corrosion environment, it is difficult for a single corrosion inhibitor to meet the requirements, so it is necessary to compound and synergize the corrosion inhibitor. The so-called synergistic effect of corrosion inhibitors refers to the phenomenon that the performance of a corrosion inhibitor is enhanced and improved due to the addition of other substances (or corrosion inhibitors). That is, in a corrosive medium, a mixed corrosion inhibitor composed of two or more corrosion-inhibiting substances will have a better corrosion-inhibiting effect. The improvement of this corrosion inhibition effect is not a simple addition of the two components but promotes each other to achieve the effect of $1 + 1 \gg 2$. At present, most of the compounding research is based on the synergistic effect of corrosion inhibition between imidazoline and other types of corrosion inhibitors, which can be divided into synergistic effect with halide ions and those containing N, S, O, P and other atoms. Most of the corrosion inhibitors that are practically used are made up of a variety of substances [54]. The use of synergy can reduce the amount of corrosion inhibitor based on ensuring the corrosion inhibition effect, or obtain a higher corrosion inhibition rate and solve the difficulties that are difficult to overcome by a single component.

The most common synergistic effect of CO_2 corrosion inhibitors is the synergistic effect of imidazoline inhibitors and thiourea. For example, Okafor [55] studied the inhibition and adsorption behavior of 2-undecyl-1-sodium ethanoate-imidazoline salt (2M2) and thiourea (TU) on N80 mild steel in CO_2-saturated 3 wt% NaCl solutions at 25 °C, 1 bar CO_2 partial pressure, and the results show that the inhibition process is attributed to the formation of an adsorbed film of 2M2 and TU via the inhibitor's polycentric adsorption sites on the metal surface which protects the metal against corrosion. Wang [43, 56] investigated the inhibition performance and inhibition mechanism of imidazoline with thioureido to prevent corrosion of steel in saltwater saturated with CO_2. The results show that the chemisorption of imidazoline and thiourea molecules on Q235 steel depends on the imidazoline ring and thiourea, and the compound corrosion inhibitor proves to has good corrosion inhibition performance. In a 3 wt% CO_2 saturated NaCl solution, Heydari [57] examined the corrosion inhibition of an amido-imidazoline derivative on X52 steel. Concentration affected the degree to which amido-imidazoline inhibited corrosion of steel, and a synergistic effect meant that amido-imidazoline's effectiveness at inhibiting corrosion was increased by the addition of iodide ions. In both the supercritical CO_2-saturated aqueous phase and the water-saturated supercritical CO_2 phase, Xiang [58] investigated the corrosion inhibition effects of imidazoline and piperazine on N80 steel. It was discovered that piperazine has an exceptional neutralizing effect and might function as a pH stabilizer. $Fe_2(SO_4)_3$ was confirmed in the product scale, and Fe_3C was present on the steel's surface. Liu [59] used the mass loss method, electrochemical testing, SEM morphological analysis, and molecular dynamics modeling to examine the synergistic inhibitory effect of imidazoline and thiourea. The outcomes demonstrated a good synergistic corrosion inhibition effect of imidazoline and thiourea. The first layer was formed by thiourea molecules adhering to the metal surface, while the second layer was created by imidazoline molecules adhering to the thiourea. The amount of water molecules that were initially adsorbed on the Fe surface is decreased, and the pace at which water molecules diffuse through the corrosion inhibitor film is slowed down. These two effects act in coordination to suppress metal corrosion.

Zhang [60] tested the synergistic effect between imidazoline-based dissymmetric bis-quaternary ammonium (DBA) salts and thiourea (TU) in the CO_2-saturated water at 80 °C. The results shown that DBA alone has poor inhibition efficiency, and TU has a narrow inhibition concentration range at high temperatures. However, as shown in Fig. 7.7, the morphological features of the Q235 steel surface after immersion in the CO_2-saturated water with DBA–TU compound inhibitor is smooth and bright, which demonstrates DBA–TU forms a complete adsorption film on the Q235 steel surface, and that DBA–TU can be used as a high-temperature corrosion inhibitor against CO_2 corrosion.

The synergistic effect of corrosion inhibition between organics and halide ions is a relatively mature corrosion inhibition synergistic effect system. The characteristic adsorption makes the metal surface negatively charged, which is conducive to the adsorption of cationic corrosion inhibitor molecules on the metal surface, thereby improving the corrosion inhibition effect. For example, Zhao [61] investigated the synergistic inhibition effect of oleic-based imidazoline and sodium benzoate on mild steel corrosion in a CO_2-saturated brine solution. The results show that oleic-based imidazoline has a certain protective effect on the CO_2 corrosion of low carbon steel, and its corrosion inhibition effect can be enhanced when used in combination with sodium benzoate. The synergistic corrosion inhibition effect of oleic-based imidazoline and sodium benzoate is obvious.

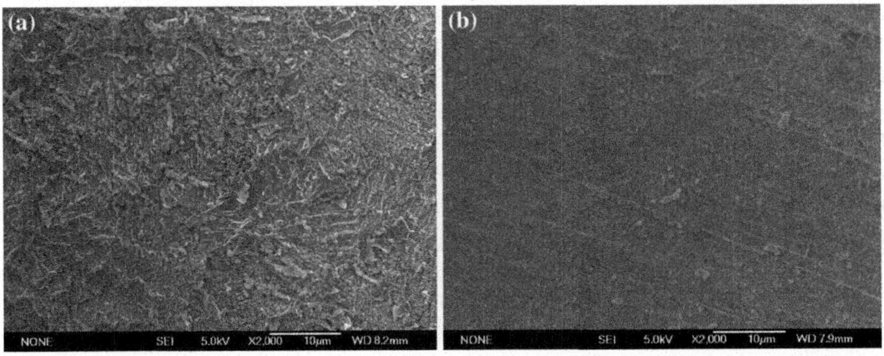

Fig. 7.7 SEM image of Q235 steel in the CO_2-saturated water in the absence and presence of the inhibitors at 80 °C **a** after immersion without inhibitor, **b** after immersion in the presence of DBA–TU [60]

7.3 Applications

Previous studies of CO_2 corrosion inhibitors have mainly focused on CO_2 corrosion problems in oil and gas pipelines, and that type of CO_2 corrosion occurs at lower CO_2 partial pressures [62–66]. As for CCUS, however, the environment and the operating conditions (i.e. pressure and temperature) are different compared with oil and gas production. There is a need to take special care when transferring the experience in oil and gas production to CCUS environment since the pressure of CCUS pipelines is about 3 times of that in oil and gas pipelines, and the measurement of corrosion kinetics is difficult under high pressure of CO_2, water, and other impurities. Therefore, the performance evaluation of corrosion inhibitors under those conditions is a tedious task. The impedance spectra at different times under supercritical CO_2 conditions are similar to those under non-supercritical conditions, indicating that there is no essential difference in the electrochemical corrosion mechanism between supercritical and non-supercritical conditions. However, the positive shift of potential and the sharp increase in polarization resistance of steel under supercritical CO_2 conditions occur earlier, so the corrosion rate is higher. Under supercritical conditions, the increase of CO_2 partial pressure increases the concentration of corrosive substances (such as H^+, HCO_3^- and H_2CO_3), which accelerates the corrosion of steel.

In the process of CO_2 capture, aqueous solutions of alkanolamines are one of the most commonly used CO_2 absorbents. Monoethanolamine (MEA), diethanolamine (DEA), and N-methyl diethanolamine (MDEA) have respective advantages in the process of CO_2 capture. The results showed that these alkanolamines at a concentration of 300 ppmv were between 55 and 67% in the aqueous phase, whereas the inhibition efficiencies were low in the supercritical CO_2 phase [67]. The results indicated that the inhibition mechanism of alkanolamines was attributed to both the neutralization and adsorption effects on the steel surface [67, 68].

As for the CO_2 transportation, Hua [69, 70] studied the corrosion of X65 steel in the supercritical CO_2-saturated water, water-saturated SC-CO_2 and under-saturated SC-CO_2 conditions phase at 80 bar and 50 °C to simulate the different conditions of CO_2 transportation in CCUS. The results showed that the X65 steel reached 4 mm/year and the pitting rate was 1.5 mm/year after 96 h of corrosion. It was found that the Fe_3C film started to form in the first 6 h and generated after 14 h. After 48 h, $FeCO_3$ began to precipitate, and the outer layer of crystalline $FeCO_3$ formed after 96 h. The results highlight the importance of determining the level of localized corrosion in the environment representing different stages of CO_2 transport in CCUS.

Besides, Cui [71] used a rotating autoclave to study the corrosion characteristics of J55 casing steel and N80 tubing steel at a pressure of 5 MPa and 50 ~ 200 °C environment containing CO_2 and NaCl, and the results showed that oil-soluble imidazoline acts as a good inhibitor for corrosion of J55 and N80 in the environment containing CO_2 and NaCl. Choi [72, 73] used electrochemical measurements and surface analytical techniques to investigate the corrosion behavior of low Cr alloy steels and the

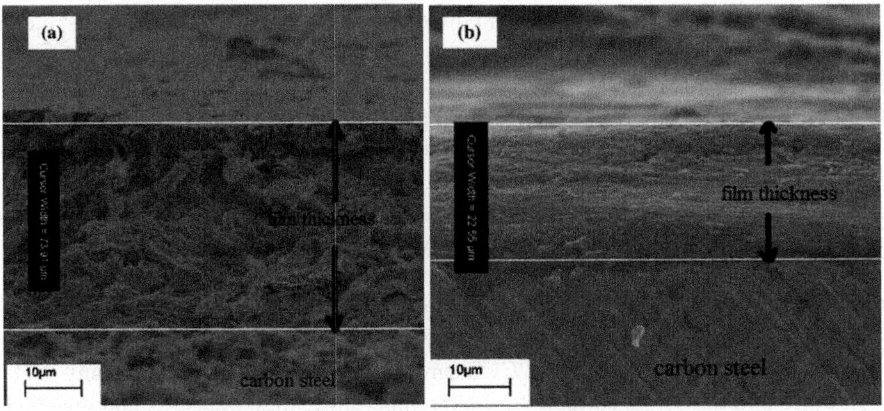

Fig. 7.8 Surface morphology of X52 steel in high-pressure CO_2 (6 MPa) saturated brine **a** without imidazoline **b** with 100 ppm of the imidazoline at 60 °C for 24 h [74]

results showed that the addition of imidazoline-based inhibitors decreased corrosion rate significantly from 90 mm/year to below 0.1 mm/year at supercritical CO_2 condition (12 MPa CO_2, 80 °C).

Under an experimental condition of 60 °C and high CO_2 pressures (10, 40, and 60 bar), Mustafa [74] investigated the impact of an imidazoline-based inhibitor on the formation, microstructure, and thickness of the corrosion product film that formed on X52 steel. The results, as shown in Fig. 7.8, demonstrated that the inhibitor had a considerable impact on the thickness and surface morphology of the corrosion product film [74]. $FeCO_3$ and Fe_3C made up the majority of the corrosion product film, which significantly slowed down corrosion to less than 2 mm/year.

This chapter introduces main CO_2 corrosion inhibitors in CCUS environment, including a general introduction to corrosion inhibitors, the CO_2 corrosion mechanism of steel and anticorrosion mechanism of corrosion inhibitors, and the classification, synthesis, and application of corrosion inhibitors. Future research directions are summarized as follows:

(1) Research on the corrosion inhibition theory and rational utilization of corrosion inhibitors under high-flow rate conditions is desired. Questions need to be answered regarding whether the corrosion inhibitors developed and screened based on experiments and theories under static or quasi-static conditions are suitable for high-flow rate, multiphase-flow conditions.
(2) The on-site corrosion inhibitor evaluation method that can easily monitor and control the corrosion status of equipment is desired. Real-time and online monitoring has become one of the development directions of corrosion inhibitor evaluation methods.
(3) Local corrosion and its protection, evaluation, and prediction, and the development of more efficient, multi-functional, and environmental-friendly corrosion inhibitors will become the focus of research [75–78]. Extracting or separating

effective corrosion inhibitor components from natural plants, medicines, etc. is also an important direction for the development of corrosion inhibitors in the future.

(4) Development of multiphase corrosion inhibitors suitable for gas/liquid/solid multiphase corrosion systems is desired. Furthermore, most of the existing research is aimed at exploring the best dosage of the corrosion inhibitor added, while the mechanism of action between the corrosion inhibitor and the CO_2 corrosion product film is less studied, which is an important direction for future research.

(5) Although the CO_2 corrosion inhibitors described above do reduce the CO_2 corrosion rate, some of them do not have good performance to inhibit the corrosion of steel casing under CCUS environment. Therefore, developing metal corrosion inhibitors suitable for high temperature, high pressure, and high CO_2 concentration under different CCUS environments is an important research direction.

References

1. Suba, P.I.V.: Corrosion Inhibition Performance of Sulfur and Nitrogen Based Environmentally Friendly Inhibitors for CO_2 Absorption Plants. The University of Regina, Saskatchewan (2020)
2. Xiang, Y., Song, C., Li, C., Yao, E., Yan, W.: Characterization of 13Cr steel corrosion in simulated EOR-CCUS environment with flue gas impurities. Process Saf. Environ. Prot. **140**, 124–136 (2020)
3. Kang, D., Lee, M.-G., Yoo, Y., Park, J.: Absorption characteristics of potassium carbonate-based solutions with rate promoters and corrosion inhibitors. J. Mater. Cycles Waste Manage. **20**(3), 1562–1573 (2018)
4. Xiang, Y., Xie, W., Ni, S., He, X.: Comparative study of A106 steel corrosion in fresh and dirty MEA solutions during the CO_2 capture process: effect of NO_3^-. Corros. Sci. **167**, 108521 (2020)
5. Barker, R., Hua, Y., Neville, A.: Internal corrosion of carbon steel pipelines for dense-phase CO_2 transport in carbon capture and storage (CCS)–a review. Int. Mater. Rev. **62**(1), 1–31 (2017)
6. Cole, I.S., Corrigan, P., Sim, S., Birbilis, N.: Corrosion of pipelines used for CO_2 transport in CCS: is it a real problem? Int. J. Greenhouse Gas Control **5**(4), 749–756 (2011)
7. Sim, S., Cole, I.S., Choi, Y.S., Birbilis, N.: A review of the protection strategies against internal corrosion for the safe transport of supercritical CO_2 via steel pipelines for CCS purposes. Int. J. Greenhouse Gas Control **29**, 185–199 (2014)
8. Choi, Y.-S., Nešić, S.: Determining the corrosive potential of CO_2 transport pipeline in high pCO_2–water environments. Int. J. Greenhouse Gas Control **5**(4), 788 (2011)
9. Choi, Y.-S., Young, D., Nešić, S., Gray, L.G.S.: Wellbore integrity and corrosion of carbon steel in CO_2 geologic storage environments: a literature review. Int. J. Greenhouse Gas Control **16**, S70 (2013)
10. Nesic, S., Postlethwaite, J., Olsen, S.: An electrochemical model for prediction of corrosion of mild steel in aqueous carbon dioxide solutions. Corrosion **52**(4), 280–294 (1996)
11. Nordsveen, M., Nesic, S., Nyborg, R., Stangeland, A.: A mechanistic model for carbon dioxide corrosion of mild steel in the presence of protective iron carbonate films-Part 1: theory and verification. Corrosion **59**(5), 443–456 (2003)

12. Cui, Z.D., Wu, S.L., Li, C.F., Zhu, S.L., Yang, X.J.: Corrosion behavior of oil tube steels under conditions of multiphase flow saturated with super-critical carbon dioxide. Mater. Lett. **58**(6), 1035–1040 (2004)
13. Cui, Z.D., Wu, S.L., Zhu, S.L., Yang, X.J.: Study on corrosion properties of pipelines in simulated produced water saturated with supercritical CO_2. Appl. Surf. Sci. **252**(6), 2368–2374 (2006)
14. Choi, Y.-S., Nešić, S.: Corrosion behavior of carbon steel in supercritical CO_2-water environments, Corrosion 2009, Atlanta, Georgia (2009)
15. Nor, A.M., Suhor, M., Mohamed, M., Singer, M.M., Nesic, S.M.: Corrosion of carbon steel in high CO_2 environment: flow effect, Corrosion 2011, Houston, Texas (2011)
16. Zhang, X., Zevenbergen, J., Benedictus, T.: Corrosion studies on casing steel in CO_2 storage environments. Energy Procedia **37**, 5816–5822 (2013)
17. Wei, L., Zhang, Y., Pang, X., Gao, K.: Corrosion behaviors of steels under supercritical CO_2 conditions. Corros. Rev. **33**(3–4), 151–174 (2015)
18. Cui, G., Yang, Z., Liu, J., Li, Z.: A comprehensive review of metal corrosion in a supercritical CO_2 environment. Int. J. Greenhouse Gas Control **90**, 102814 (2019)
19. Dong, B., Zeng, D., Yu, Z., Cai, L., Yu, H., Shi, S., Tian, G., Yi, Y.: Major corrosion influence factors analysis in the production well of CO_2 flooding and the optimization of relative anti-corrosion measures. J. Pet. Sci. Eng. **200**, 108052 (2021)
20. Li, C., Xiang, Y., Li, W.: Initial corrosion mechanism for API 5L X80 steel in CO_2/SO_2-saturated aqueous solution within a CCUS system: Inhibition effect of SO_2 impurity. Electrochim. Acta **321**, 134663 (2019)
21. Zhang, G.A., Liu, D., Li, Y.Z., Guo, X.P.: Corrosion behaviour of N80 carbon steel in formation water under dynamic supercritical CO_2 condition. Corros. Sci. **120**, 107–120 (2017)
22. Tang, Y., Guo, X.P., Zhang, G.A.: Corrosion behaviour of X65 carbon steel in supercritical-CO_2 containing H_2O and O_2 in carbon capture and storage (CCS) technology. Corros. Sci. **118**, 118–128 (2017)
23. Jiang, X., Zheng, Y.G., Ke, W.: Corrosion inhibitor performances for carbon dioxide corrosion of N80 steel under static and flowing conditions. Corrosion **61**(4), 326–334 (2005)
24. Jiang, X., Zheng, Y.G., Ke, W.: Effect of flow velocity and entrained sand on inhibition performances of two inhibitors for CO_2 corrosion of N80 steel in 3% NaCl solution. Corros. Sci. **47**(11), 2636–2658 (2005)
25. Jiang, X., Zheng, Y.G., Qu, D.R., Ke, W.: Effect of calcium ions on pitting corrosion and inhibition performance in CO_2 corrosion of N80 steel. Corros. Sci. **48**(10), 3091–3108 (2006)
26. Choi, Y.-S., Nesic, S., Young, D.: Effect of impurities on the corrosion behavior of CO_2 transmission pipeline steel in supercritical CO_2−water environments. Environ. Sci. Technol. **44**(23), 9233–9238 (2010)
27. Yue, X., Zhao, M., Zhang, L., Zhang, H., Li, D., Lu, M.: Correlation between electrochemical properties and stress corrosion cracking of super 13Cr under an HTHP CO_2 environment. RSC Adv. **8**(43), 24679–24689 (2018)
28. Usman, B.J., Ali, S.A.: Carbon dioxide corrosion inhibitors: a review. Arabian J. Sci. Eng. **43**(1), 1–22 (2018)
29. Hua, Y., Shamsa, A., Barker, R., Neville, A.: Protectiveness, morphology and composition of corrosion products formed on carbon steel in the presence of Cl^-, Ca^{2+} and Mg^{2+} in high pressure CO_2 environments. Appl. Surf. Sci. **455**, 667–682 (2018)
30. Han, J., Carey, J.W., Zhang, J.: Effect of sodium chloride on corrosion of mild steel in CO_2-saturated brines. J. Appl. Electrochem. **41**(6), 741–749 (2011)
31. Cerruti, S.: An overview of corrosion resistant alloy steel selection and requirements for oil and gas industry, Conference: Convegno IGF XIV Trento, 14 (1998)
32. Sorensen, P.A., Kiil, S., Dam-Johansen, K., Weinell, C.E.: Anticorrosive coatings: a review. J. Coat. Technol. Res. **6**(2), 135–176 (2009)
33. Singha, A., Quraishi, M.A.: Acidizing corrosion inhibitors: a review. J. Mater. Environ. Sci. **6**(1), 224–235 (2015)

34. Kuznetsov, Y.I.: Current state of the theory of metal corrosion inhibition. Prot. Met. **38**(2), 103–111 (2002)
35. Chauhan, D.S., Quraishi, M.A., Sorour, A.A., Verma, C.: A review on corrosion inhibitors for high-pressure supercritical CO_2 environment: challenges and opportunities. J. Pet. Sci. Eng. **215**, 110695 (2022)
36. Hua, Y., Barker, R., Charpentier, T., Ward, M., Neville, A.: Relating iron carbonate morphology to corrosion characteristics for water-saturated supercritical CO_2 systems. J. Supercrit. Fluids **98**, 183–193 (2015)
37. Li, T., Yang, Y., Gao, K., Lu, M.: Mechanism of protective film formation during CO_2 corrosion of X65 pipeline steel. Journal of University of Science and Technology Beijing, Mineral, Metallurgy, Material **15**(6), 702–706 (2008)
38. Dariva, C.G., Galio, A.F.: Corrosion inhibitors–principles, mechanisms and applications, developments in corrosion protection, Chap. 16, pp. 365–378, 1st edn. InTech (2014)
39. Liu, F.G., Du, M., Zhang, J., Qiu, M.: Electrochemical behavior of Q235 steel in saltwater saturated with carbon dioxide based on new imidazoline derivative inhibitor. Corros. Sci. **51**(1), 102–109 (2009)
40. Wang, Z.Z., Li, Y.Y., Zhang, G.A.: Inhibitive effects of inhibitors on the galvanic corrosion between N80 carbon steel and 13Cr stainless steel under dynamic supercritical CO_2 conditions. Corros. Sci. **146**, 121–133 (2019)
41. Wei, L., Chen, Z., Guo, X.: Inhibition behavior of an imidazoline inhibitor for carbon steel in a supercritical CO_2/H_2O system. J. Electrochem. Soc. **164**(9), C602–C609 (2017)
42. Mohamed, M.F.: Water Chemistry and Corrosion Inhibition in High Pressure CO_2 Corrosion of Mild Steel. Ohio University, USA (2015)
43. Wang, B., Du, M., Zhang, J., Gao, C.J.: Electrochemical and surface analysis studies on corrosion inhibition of Q235 steel by imidazoline derivative against CO_2 corrosion. Corros. Sci. **53**(1), 353–361 (2011)
44. Farelas, F., Ramirez, A.: Carbon dioxide corrosion inhibition of carbon steels through bis-imidazoline and imidazoline compounds studied by EIS. Int. J. Electrochem. Sci. **5**(6), 797–814 (2010)
45. Okafor, P.C., Liu, X., Zheng, Y.G.: Corrosion inhibition of mild steel by ethylamino imidazoline derivative in CO_2-saturated solution. Corros. Sci. **51**(4), 761–768 (2009)
46. Zhang, G., Chen, C., Lu, M., Chai, C., Wu, Y.: Evaluation of inhibition efficiency of an imidazoline derivative in CO_2-containing aqueous solution. Mater. Chem. Phys. **105**(2–3), 331–340 (2007)
47. Lopez, D.A., Simison, S.N., de Sanchez, S.R.: Inhibitors performance in CO_2 corrosion EIS studies on the interaction between their molecular structure and steel microstructure. Corros. Sci. **47**(3), 735–755 (2005)
48. Jovancicevic, V., Ramachandran, S., Prince, P.: Inhibition of carbon dioxide corrosion of mild steel by imidazolines and their precursors. Corrosion **55**(5), 449–455 (1999)
49. Zhang, C., Cui, F., Zeng, G., Jiang, M., Yang, Z., Yu, Z., Zhu, M., Shen, L.: Quaternary ammonium compounds (QACs): a review on occurrence, fate and toxicity in the environment. Sci. Total Environ. **518–519**, 352–362 (2015)
50. Chauhan, D.S., Quraishi, M.A., Mazumder, M.A.J., Ali, S.A., Aljeaban, N.A., Alharbi, B.G.: Design and synthesis of a novel corrosion inhibitor embedded with quaternary ammonium, amide and amine motifs for protection of carbon steel in 1 M HCl, J. Mol. Liq. **317** (2020)
51. Zhao, J., Duan, H., Jiang, R.: Synergistic corrosion inhibition effect of quinoline quaternary ammonium salt and Gemini surfactant in H_2S and CO_2 saturated brine solution. Corros. Sci. **91**, 108–119 (2015)
52. Khodyrev, Y.P., Batyeva, E.S., Badeeva, E.K., Platova, E.V., Tiwari, L., Sinyashin, O.G.: The inhibition action of ammonium salts of O, O′-dialkyldithiophosphoric acid on carbon dioxide corrosion of mild steel. Corros. Sci. **53**(3), 976–983 (2011)
53. Desimone, M.P., Grundmeier, G., Gordillo, G., Simison, S.N.: Amphiphilic amido-amine as an effective corrosion inhibitor for mild steel exposed to CO_2 saturated solution: polarization, EIS and PM-IRRAS studies. Electrochim. Acta **56**(8), 2990–2998 (2011)

54. Zhang, J., Zhang, Q., Ren, H., Zhao, W., Zhang, H.: Inhibition performance of 2-mercaptobenzothiazole derivatives in CO_2 saturated solution and its adsorption behavior at Fe surface. Appl. Surf. Sci. **253**(18), 7416–7422 (2007)
55. Okafor, P.C., Liu, C.B., Liu, X., Zheng, Y.G., Wang, F., Liu, C.Y., Wang, F.: Corrosion inhibition and adsorption behavior of imidazoline salt on N80 carbon steel in CO_2-saturated solutions and its synergism with thiourea. J. Solid State Electrochem. **14**(8), 1367–1376 (2010)
56. Wang, B., Du, M., Zhang, J.: Study of the inhibition mechanism of imidazoline derivative inhibitor on CO_2 corrosion for Q235 steel. Advanced Materials Research, Trans Tech Publ 981–984 (2009)
57. Heydari, M., Javidi, M.: Corrosion inhibition and adsorption behaviour of an amido-imidazoline derivative on API 5L X52 steel in CO_2-saturated solution and synergistic effect of iodide ions. Corros. Sci. **61**, 148–155 (2012)
58. Xiang, Y., Long, Z., Li, C., Huang, H., He, X.: Inhibition of N80 steel corrosion in impure supercritical CO_2 and CO_2-saturated aqueous phases by using imino inhibitors. Int. J. Greenhouse Gas Control **63**, 141–149 (2017)
59. Liu, J., Gao, G., Fang, X., Wang, L., Tang, S., Xing, X., Li, Z., Cui, G., He, G., Wang, Z.: Synergistic inhibition effect of imidazoline and thiourea: evidence from experiments and molecular dynamics simulation. Corrosion **76**(12), 1194–1206 (2020)
60. Zhang, J., Sun, X., Ren, Y., Du, M.: The synergistic effect between imidazoline-based dissymmetric bis-quaternary ammonium salts and thiourea against CO_2 corrosion at high temperature. J. Surfactants Deterg. **18**(6), 981–987 (2015)
61. Zhao, J., Chen, G.: The synergistic inhibition effect of oleic-based imidazoline and sodium benzoate on mild steel corrosion in a CO_2-saturated brine solution. Electrochim. Acta **69**, 247–255 (2012)
62. Askari, M., Aliofkhazraei, M., Jafari, R., Hamghalam, P., Hajizadeh, A.: Downhole corrosion inhibitors for oil and gas production—a review. Appl. Surf. Sci. **6**, 100128 (2021)
63. Li, J., Wang, C., Liu, J., Liu, L., Yin, Z., Zhang, Y., Li, H., He, F.: Corrosion analysis, and use of an inhibitor in oil wells. Res. Chem. Intermed. **40**(2), 649–660 (2014)
64. Finšgar, M., Jackson, J.: Application of corrosion inhibitors for steels in acidic media for the oil and gas industry: a review. Corros. Sci. **86**, 17–41 (2014)
65. Nam, N.D., Bui, Q.V., Mathesh, M., Tan, M.Y.J., Forsyth, M.: A study of 4-carboxyphenylboronic acid as a corrosion inhibitor for steel in carbon dioxide containing environments. Corros. Sci. **76**, 257–266 (2013)
66. Rajeev, P., Surendranathan, A.O., Murthy, C.S.N.: Corrosion mitigation of the oil well steels using organic inhibitors—a review. J. Mater. Environ. Sci. **3**(5), 856–869 (2012)
67. Xiang, Y., Long, Z., Li, C., Yan, W.: Neutralization and adsorption effects of various alkanolamines on the corrosion behavior of N80 steel in supercritical CO_2 with impurities. Corrosion **75**(8), 999–1011 (2019)
68. Xiang, Y., Huang, H., Long, Z., Li, C., Yan, W.: Role of residual 2-amino-2-methyl-1-propanol and piperazine in the corrosion of X80 steel within an impure supercritical CO_2 environment as relevant to CCUS. Int. J. Greenhouse Gas Control **82**, 127–137 (2019)
69. Hua, Y., Barker, R., Neville, A.: Comparison of corrosion behaviour for X-65 carbon steel in supercritical CO_2-saturated water and water-saturated/unsaturated supercritical CO_2. J. Supercrit. Fluids **97**, 224–237 (2015)
70. Hua, Y., Barker, R., Neville, A.: Effect of temperature on the critical water content for general and localised corrosion of X65 carbon steel in the transport of supercritical CO_2. Int. J. Greenhouse Gas Control **31**, 48–60 (2014)
71. Cui, L., Kang, W., You, H., Cheng, J., Li, Z.: Experimental study on corrosion of J55 casing steel and N80 tubing steel in high pressure and high temperature solution containing CO_2 and NaCl. Tribol. Int. **7**(1), 13 (2020)
72. Choi, Y.-S., Hassani, S., Vu, T.N., Nešić, S., B. Abas, A.Z., Nor, A.M., Suhor, M.F.: Strategies for corrosion inhibition of carbon steel pipelines under supercritical CO_2/H_2S environments. Corrosion **75**(10), 1156–1172 (2019)

73. Choi, Y.-S., Hassani, S., Nam Vu, T., Nesic, S., B Abas, A.Z., Nor, A.M., Suhor, M.F.: Corrosion inhibition of pipeline steels under supercritical CO_2 environment, Corrosion 2017, New Orleans, Louisiana, USA (2017)
74. Mustafa, A.H., Ari-Wahjoedi, B., Ismail, M.C.: Inhibition of CO_2 corrosion of X52 steel by imidazoline-based inhibitor in high pressure CO_2-water environment. J. Mater. Eng. Perform. **22**(6), 1748–1755 (2013)
75. Marciales, A., Haile, T., Ahvazi, B., Ngo, T.D., Wolodko, J.: Performance of green corrosion inhibitors from biomass in acidic media. Corros. Rev. **36**(3), 239–266 (2018)
76. Hossain, N., Asaduzzaman Chowdhury, M., Kchaou, M.: An overview of green corrosion inhibitors for sustainable and environment friendly industrial development. J. Adhes. Sci. Technol. **35**(7), 673–690 (2021)
77. Singh, A., Lin, Y., Ebenso, E.E., Liu, W., Pan, J., Huang, B.: Gingko biloba fruit extract as an eco-friendly corrosion inhibitor for J55 steel in CO_2 saturated 3.5% NaCl solution. Ind. Eng. Chem. **24**, 219–228 (2015)
78. Chauhan, D.S., Quraishi, M.A., Qurashi, A.: Recent trends in environmentally sustainable sweet corrosion inhibitors. J. Mol. Liq. **326**, 115117 (2021)

The manufacturer's authorised representative in the EU is Springer Nature Customer Service Centre GmbH, Europaplatz 3, 69115 Heidelberg, Germany. If you have any concerns regarding our products, please contact ProductSafety@springernature.com

Printed and bound by CPI Group (UK) Ltd, Croydon, CR0 4YY

01/10/2026

02225877-0003